Optical Fibers

Optical Fibers

TAKANORI OKOSHI

Department of Electronic Engineering
University of Tokyo
Tokyo, Japan

 1982

ACADEMIC PRESS

A Subsidiary of Harcourt Brace Jovanovich, Publishers

New York   London
Paris   San Diego   San Francisco   São Paulo   Sydney   Tokyo   Toronto

7190 - 1747

CHEMISTRY

ACADEMIC PRESS, INC.
111 Fifth Avenue, New York, New York 10003

United Kingdom Edition published by
ACADEMIC PRESS, INC. (LONDON) LTD.
24/28 Oval Road, London NW1 7DX

Library of Congress Cataloging in Publication Data

Okoshi, Takanori.
   Optical fibers.

   Includes index.
   1. Fiber optics. 2. Optical communications.
I. Title. [DNLM: 1. Fiber optics. QC 448 O41o]
TA1800.O38   621.36'92   81-17594
ISBN 0-12-525260-9   AACR2

This translation copyright © 1982, by Takanori Okoshi. Original Japanese-language edition, Hikari-Faiba no Kiso, copyright © 1977, by Takanori Okoshi, Katsunari Okamoto, and Kazuo Hotate, published by Ohm-sha, Tokyo, Japan. English translation rights arranged through Japan UNI Agency, Inc.

PRINTED IN THE UNITED STATES OF AMERICA

82 83 84 85   9 8 7 6 5 4 3 2 1

Contents

TA1800
03813
1982
ChEM

8 Optical Fibers Having Structural Fluctuations

9 Measurement of Refractive-Index Profile of Optical Fibers

10 Measurement of Transmission Characteristics of Optical Fibers

11 Concluding Remarks

Appendix

Preface

The advent of low-loss optical fibers in the early 1970s was a major event in the field of communications in the third quarter of the twentieth century. Since then their design and fabrication techniques have shown exceptionally rapid progress. Engineers now believe that in the distant future the optical fiber will change dramatically the entire aspect of communications in human society.

This book describes the theoretical basis (electromagnetic theories) of transmission characteristics of optical fibers. The material and fabrication technologies are not described systematically, but only briefly where necessary to assist comprehension of the transmission characteristics. The reason for such a restriction is simple: the material and fabrication technologies are still progressing rapidly, whereas the electromagnetic theories have reached in the past several years a level that now enables us to give a textbook-like description on the entire field.

This book is entitled simply "Optical Fibers." If a more lengthy but descriptive title were preferable, it would be "Analysis and Design of Optical Fibers," because most of the description and discussion in this book are motivated by the following two basic questions:

(1) How can we analyze the transmission characteristics (more specifically, the group-delay characteristics, or in other words the modulation-frequency response or the impulse response) of an optical fiber having an arbitrary refractive-index profile?

(2) What is the optimum refractive-index profile with which the group-delay spread of an optical fiber is minimized?

The author never believes, of course, that the readers of this book are concerned only with the analysis of arbitrary-profile fibers or the optimum design of the refractive-index profile. Some readers may be interested mainly in the characteristics of uniform-core fibers, or mainly in the measurement techniques, or even only in the historical aspects of the optical-fiber techniques. To meet such various possible requirements, the author tried to make this a book of as general and comprehensive a nature as possible, still keeping the unity of all the chapters and sections by reminding the readers (and also the author himself) of the above two questions at various steps of the description.

The author and his colleagues started research on optical fibers in 1972. At that time the theories on transmission characteristics of optical fibers had not matured; the relations among different kinds of theories were not well understood, and even worse, different terminologies and symbols were used in different kinds of theories. As the result of the vigorous investigations in the early 1970s, the relations among different theories are now fairly well understood. However, the confusion in the customarily used symbols remains, often annoying investigators and students who study and compare those theories. That was one of the reasons the author and his colleagues wrote the original Japanese edition of this book. It is also the reason the author felt compelled to translate it into English.

The original Japanese edition of this book, "Hikari Faiba no Kiso" (Fundamentals of Optical Fibers), was written jointly by T. Okoshi, K. Okamoto, and K. Hotate, edited by T. Okoshi, and published by the Ohm Publishing Company, Tokyo, in July 1977. The translation of the Japanese text into English began shortly before the publication of the Japanese edition, and was completed in September 1979. In the course of the translation, the author found some chapters and sections unsatisfactory; therefore, he rewrote and supplemented those and reorganized the entire volume.

A bird's-eye view of the construction of this book will be found in Chapter 1, Section 1.4, of the text. For the convenience of readers who have access to both the Japanese and English editions, major revisions will be mentioned in the following. The first half of Chapter 4 has fully been rewritten; the simplest mode theory for a uniform-core fiber is newly added to make the description more comprehensive; Chapter 5 has been thoroughly rewritten and new sections on the power-series expansion method and finite-element method analyses have been added. The latter half of Chapter 8 has been deleted for space restrictions. Chapter 11 has been updated. Appendixes 2A.1, 3A.1, 3A.2, 3A.4, 3A.5, 3A.6, 4A.1, 4A.2, 5A.4, 5A.9, 5A.10, 5A.11, and 9A.1 have been added.

Acknowledgments

First, I should like to thank the two coauthors of the original Japanese edition of this book, Dr. K. Okamoto, now with Ibaraki Electrical Communication Laboratory, Nippon Telegraph and Telephone Public Corporation (NTT), and Dr. K. Hotate, a lecturer of the Faculty of Engineering, University of Tokyo.

When the manuscript of the Japanese edition was prepared in 1976–1977, both of them were graduate-school Ph.D. students (probably the brightest that I have ever supervised) working on optical fibers in my laboratory. They generously permitted me to translate into English and reorganize the original Japanese edition and to publish the English edition abroad.

I am also indebted to a large number of people for the work on which this book is based. This work was performed with continuous encouragement and support from Professors S. Okamura and H. Yanai, both of the Department of Electronic Engineering, University of Tokyo. Continuous technical support has also been given by Dr. N. Inagaki and others of Ibaraki Electrical Communication Laboratory, NTT; Dr. T. Nakahara and others of Sumitomo Electric Industries, Ltd.; Dr. H. Murata and others of Furukawa Electric Co., Ltd.; and Dr. S. Tanaka and others of Fujikura Cable Works, Ltd. There are also many people from my laboratory at the University of Tokyo who helped me to prepare the figures used and to study the theory of optical fibers by introducing new papers at the weekly seminars of our laboratory.

Writing a book is really a tedious task, especially when the language used is not the author's working language. English was once my working language, but only for a year and half in 1963–1964, when I worked at Bell Laboratories, Murray Hill, New Jersey, on a leave of absence from the University of Tokyo. If the English of this book is a little better than average Japanese English, I owe that difference mostly to my supervisor at Bell Laboratories, Dr. J. W. Gewartowski, who taught me how to write English by correcting carefully all the details of my more than ten technical papers. Moreover, many thanks are due to Professor J. K. Butler of Southern Methodist University, who carefully worked through the entire manuscript to ensure the clarity of language and presentation in English.

Finally, I wish to express my hearty thanks to my secretary, Miss M. Onozuka, who typed the manuscript and its revisions repeatedly. Without her devoted assistance this book would never have been completed.

Optical Fibers

1 | Introduction

The history of optical communications rather than that of optical fibers is reviewed. In particular, various optical waveguiding schemes proposed before the advent of practical optical fibers are discussed. Finally, the purpose and organization of this book are stated.

1.1 Technical Background

As early as the 1940s engineers and scientists began to consider that telecommunications in the distant future would be through optical channels. At first such an idea was merely a dream originating from the fact that, since Marconi's invention of wireless telegraphy, radio engineers have continuously pursued technologies of shorter wavelengths. After 1950, however, microwave communications became practical and the definite advantages of using shorter wavelengths were demonstrated. Thus, it became a common belief that the technological push toward shorter wavelengths would reach the optical region.

Nevertheless, until the late 1960s when practical optical fibers appeared, this "common belief" had not been accompanied by a realistic system concept. In reality, it was often asserted that the development of optical communications was inevitable "because it would offer much wider bandwidth per channel than microwave or millimeter-wave channels do since the frequency itself is much higher." Today, this statement is no longer realistic, as seen in the following.

At present, the optical fiber is the only practical waveguide for optical communications. The maximum optical fiber bandwidth imaginable at present is at most several gigahertz, which is much narrower than that of a millimeter waveguide. Optical fiber communications is now being vigorously

1

investigated not because it offers a wide frequency band, but mainly because it offers a wide frequency band at economical costs when many fibers in a bundle are used, thus taking advantage of its flexibility, small diameter, and low cost.

This brief history of optical communications indicates that the advent of practical fibers in the 1960s was really an epochal event in the history of optical communications. It gave the dream of optical communications a realistic system concept and a definite technical target. We will investigate the history of optical communications and illustrate how the optical fiber appeared after various proposals of optical waveguiding schemes.

1.2   Optical Communications Before the Advent of Optical Fibers

In a broad sense, optical communication has a very long history. Even a history of its "practical uses" existed before the history of its research and development in the present century.

1.2.1   First Period—Communication by Eye

The first telecommunication scheme that appeared in the history of man was the most primitive optical communication, signal fire, which has been used by man for thousands of years.

An epoch was marked when Chappe of France invented a new telecommunication scheme, often called the "Semaphore," in 1791 [1]. In his scheme, towers equipped with privoted pointing arms for displaying signals as shown in Fig. 1.1 were constructed at appropriate intervals and manned by operators. Each operator moved the pointing arms in the same manner as he observed on the display of the neighboring tower, and thus messages were transmitted. This was the first high-speed telecommunication system in the history of man, and it had a great impact on European society. The conquest and rule of European countries by Napoleon is said to be in part due to the use of this "optical" communication system.

Chappe's optical communication system disappeared abruptly after the invention of telegraphy by Morse in 1835. Although "optical communications by eye" such as signal flags are still used, their presence is insignificant compared with modern electronic telecommunications. The history of optical communications was once interrupted by the invention of telegraphy in 1835; the history since then is not the history of optical communications but that of the research and development toward its revival.

Fig. 1.1 Chappe's communication device (reproduced with permission from "From Semaphore to Satellite" [39]).

1.2.2   Second Period—Optoelectronic Receivers

Before and during World War II, optical communications with higher information-transmission rates using parabolas and phototubes in receivers was investigated for military purposes. However, such systems never became practical before the advent of the laser which was the first light source providing sufficient collimation for such light-transmission schemes.

1.2.3   Third Period—Optical Waveguide Development

The third period began with the advent of lasers in the early 1960s. Because classical microwave technology based on electron tubes and waveguide components was just being completed at that time, many microwave engineers were seeking new challenging tasks. Some went to the fields of microwave semiconductor devices and integrated circuits, and some to the optics field including optical waveguides. As a result, this third period is characterized by their endeavors toward a really practical optical waveguide. Various waveguides such as the hollow metallic waveguide, the thin film waveguide, the lens waveguide, the mirror waveguide, and the gas-lens waveguide were proposed and investigated; however, all of them finally disappeared, as will be described. We should note that preliminary researches on the optical fiber waveguide were also performed in this period.

Various approaches proposed before the advent of practical optical fiber will be reviewed [2]. The history of optical fibers will be described separately later.

A. Iris Waveguide

The simplest among various proposals was the iris waveguide discussed theoretically by Goubau [3] in 1959. As shown in Fig. 1.2, an iris waveguide consists of many optical irises (apertures) placed at equal intervals. Because light energy at the beam edges is repeatedly removed by the irises, the equilibrium state is reached with the fundamental (TEM_{00}) Gaussian mode which has the least loss resulting from the irises.

The theoretical transmission loss of the fundamental mode at the equilibrium state is shown in Fig. 1.3 (curve a) [2]. The abscissa denotes a normalized iris area $S/\lambda d$, where S is the iris area, λ the wavelength of light, and d the iris spacing. For example, when the iris diameter is 3.4 cm, $\lambda = 1$ μm, and $d = 5$ cm, the theoretical transmission loss is as low as 1 dB/km. The validity of this theory has been proved by millimeter-wave simulation at $\lambda = 8.6$ mm by Mink [4].

However, it has also been shown by Mink's experiment that this system requires very high accuracy in positioning irises when the wavelength is short. Moreover, the light path can never be bent without the use of a mirror

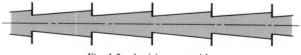

Fig. 1.2   An iris waveguide.

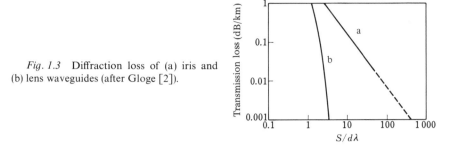

Fig. 1.3   Diffraction loss of (a) iris and (b) lens waveguides (after Gloge [2]).

or prism. Therefore, although theoretically interesting, the iris waveguide is useless for practical applications.

B. Lens and Mirror Waveguides

In an iris waveguide, an incident plane wave gradually diverges due to diffraction. In other words, the wavefront gradually curves. To periodically compensate such wavefront curvature, in 1961 Goubau proposed the structures shown in Figs. 1.4 and 1.5, the lens waveguide and the mirror waveguide [5]. As will be described, such structures have the additional advantage that the light path can be curved by shifting the lenses in the direction normal to the light path or by tilting the mirrors.

When the lens focal length is f, the optimum lens spacing is $d = 2f$. The theoretical transmission loss due to diffraction under this condition is shown in Fig. 1.3 (curve b) [2]. As shown in the figure, the diffraction loss is much lower than that of the iris waveguide, especially when $S/\lambda d$ is relatively large; the loss decreases abruptly as S increases. It is estimated that when S is several times greater than λd, the diffraction loss may be neglected in comparison with reflection loss at lens surfaces, the latter becoming the limiting factor of the transmission loss. We may reduce the reflection loss

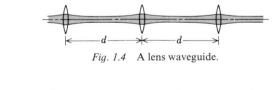

Fig. 1.4   A lens waveguide.

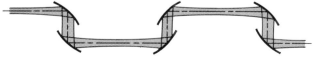

Fig. 1.5   A mirror waveguide.

to 0.5% (0.02 dB) per surface by using the best antireflection coating technique available. Therefore, when $d \simeq 200$ m the loss reduction target is 0.2 dB/km.

In 1964–1968, several lens and mirror waveguide experiments were reported. An example [6] will be described briefly. The waveguide tested was 840 m long, and consisted of 6 lenses with $f = 70$ m, a 60-mm diameter, and 0.5% reflection loss. The entire system was accommodated in a 90-mm pipeline buried 1.5 m beneath the ground to simulate an actual communication channel. A 99%-reflection semitransparent mirror and a full-reflection mirror were placed at both ends of the waveguide, and the so-called "shuttle-pulse" scheme was employed to measure the transmission characteristics for a distance of 120 km (900 lenses).

The important results follow.

(1)   At $\lambda = 632.8$ nm the transmission loss was about 0.4 dB/km.

(2)   The beam profile remained "Gaussian" during the 120-km transmission.

(3)   It was difficult to remove completely the refraction effects due to temperature gradient in the pipe, ground vibration, and the mechanical instability of the system.

Results (1) and (2) gave an optimistic prospect, but result (3) gave a pessimistic one. To overcome this difficulty, various automatic ray-stabilizing schemes were proposed in which the optical path deviation at a lens is detected and fed back to the transverse displacement of the preceding lens, so that the path deviation is minimized. However, such schemes cannot be effective for fast vibrations such as those encountered in an earthquake. Hence, in the late 1960s researchers became pessimistic about the feasibility of lens and mirror waveguides. In the arly 1970s when efforts were concentrated on optical fibers, investigations of lens and mirror waveguides naturally terminated.

To explain the background of the advent of the gas-lens waveguide, another drawback of lens and mirror waveguides should be stated: bend formation for the lens waveguide is not easy. One method is to shift the lenses in the transverse direction as shown in Fig. 1.6 and make use of the prism property of the lens in addition to its focusing property. A simple computation shows that the radius of curvature R of the bend is given in

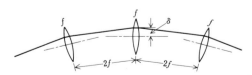

Fig. 1.6   Bending of optical path by displacement of lenses.

terms of the transverse displacement δ and the focal length f as

$$R = 2f^2/\delta. \tag{1.1}$$

For example, when $R = 500$ m and $\delta = 1$ mm, f must be as short as 50 cm. This means that 1000 lenses are needed for 1 km; hence, the reflection loss only exceeds 40 dB/km, far from practical value. If we use a mirror waveguide to avoid this difficulty, we encounter another serious problem, i.e., long-term deterioration of the mirror surface. Thus, researchers were forced to seek another scheme to bend the optical path; the gas lens seemed the most promising answer.

C. Gas-Lens Waveguide

When gas flows in a heated pipe, the gas temperature rises near the inner surface of the pipe, whereas it remains relatively low at the center. Thus, we may obtain a parabolic refractive-index distribution in which the index is higher at the axis; such an index distribution collimates a light beam around the axis, and when the pipe is gradually curved, the beam follows the axis [7, 8].

However, the preceding statement holds only for a short gas-flow range; after the gas flows a certain distance, the temperature gradient, and hence the collimating property, disappears. Moreover, in most cases, another trouble due to axially asymmetrical temperature distribution produced by gravity arises before the temperature gradient fully disappears [9]. To overcome both of these difficulties, we must remove the gas before it flows over a limited range within which the effect of gravity is not noticeable.

Figure 1.7 shows an example of a unit gas lens designed on the foregoing principle [2]. Gas flows in at both ends of this unit through the porous tube and urethane foam and is removed at the center after it flows through heated pipes. It is reported that a 40-cm focal length has been obtained by using such a unit with a wall temperature of 60°C. The beam stability for

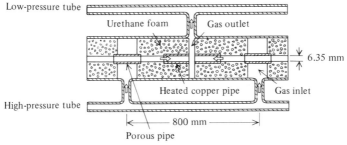

Fig. 1.7 An example of a single section of a gas-lens waveguide (after Gloge [2]).

a curved array of such gas-lens units was also demonstrated experimentally [10]. Another earlier experiment showed that a radius of curvature of 600 m was easily obtainable; the loss increase was not appreciable [11].

It was thus found that a practical bend waveguide could be obtained by the gas-lens scheme. Late in the 1960s, researchers investigated the possibility of an optical transmission system consisting of lens waveguides for straight portions and gas-lens waveguides for bent portions. However, after 1970, efforts toward such a target suddenly diminished due to the advent of a more promising counterpart, the optical fiber.

An interesting version of the gas lens is the quadrupole type proposed by Suematsu [12]. This lens consists of two high-temperature and two low-temperature rods arranged in a quadrupole. The temperature gradient produced between these rods results in a focusing property in the direction between the two high-temperature rods. A focusing waveguide can be obtained by arranging such units with 90° rotation between neighboring units. In an experiment, an equilibrium beam radius of 1 mm was achieved with 6-mm rod spacing and 12°C temperature difference [13].

D. Thin-Film Waveguide

The basic idea underlying the use of a thin film for focusing purposes follows [14].

We should first note that in the middle 1960s the dielectric loss (tan δ) of glass at optical wavelengths was of the order of magnitude of 10^{-8}, and that people never considered that this value would be dramatically improved in the near future. The tan δ of 10^{-8} means a dielectric transmission loss of about 500 dB/km. Hence, it was considered that a practical solution would be to use a very thin film as the guide of the "surface wave" so that a rather small part of the transmitted energy passes through the lossy guiding medium.

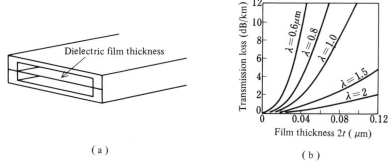

(a)   (b)

Fig. 1.8   The microguide: (a) structure, (b) transmission loss (computed).

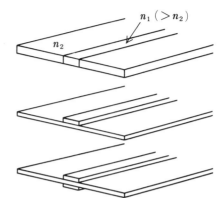

$n_1 \, (> n_2)$

n_2

Fig. 1.9   Transverse focusing schemes in thin-film waveguide.

Karbowiak proposed a structure shown in Fig. 1.8a, in which a dielectric thin film is tightly supported by a rectangular frame, and named it "microguide" [14]. Figure 1.8b shows the theoretical loss of such a structure; the refractive index of the film $n = 1.4$ and loss angle $\tan \delta = 10^{-8}$ are assumed. It is found that as the film thickness $2t$ decreases, the transmission loss is reduced; e.g., when $\lambda = 1 \, \mu$m and $2t = 0.02 \, \mu$m, the loss is as low as 2 dB/km.

This loss estimation seems encouraging, but we must note that the assumed film thickness is not practical, and that only a very weak focusing property may be expected in the direction normal to the film and no focusing in the direction parallel to it. Later, Kumagai *et al.* [15] proposed various focusing schemes in the direction parallel to the film as shown in Fig. 1.9 and also investigated loss due to the bend in the direction normal to the film [16].

E. Hollow Metallic Waveguide

Late in the 1950s, researchers were considering the hollow metallic optical waveguide [17]. Due to the skin-effect phenomena, the surface resistance of metal is proportional to $\lambda^{-1/2}$ when the surface is ideally flat, where λ denotes wavelength. Therefore, the loss constant of a metallic waveguide is

$$\alpha \propto \lambda^{3/2} S^{-3}, \tag{1.2}$$

where S is the cross-sectional area of the waveguide. For example, when a light beam with a 1-μm wavelength is transmitted through a 0.5-mm-diameter copper pipe in the TE_{01} mode, the theoretical transmission loss is 1.8 dB/km [18]. However, in practice, such a low transmission loss can never be obtained because of the following problems.

(1)   Single-mode transmission is practically impossible. For high-order modes the loss increases abruptly; e.g., 6.05 dB/km for the TE_{02} mode.

(2)   The inner surface of the tube can never be polished to optical flatness.

(3)   The loss increases abruptly when the waveguide bends only slightly.

(4)   Equation (1.2) does not hold precisely up to optical frequencies because the material constants vary.

Therefore, discussions of the feasibility of such a waveguide suddenly terminated after theoretical studies in the 1960s.

A typical example of the experiments performed late in the 1960s will be described [19]. The inner surface of a straight glass pipe with a 2-cm inner diameter and 100-m length was coated by aluminum with 10-μm flatness. The measured transmission loss at $\lambda = 632.8$ nm was 57 dB/km for an incident beam angle of 0.2°.

F. Hollow Dielectric Waveguide

A hollow dielectric pipe may also be used as an optical waveguide. This possibility, long known, was newly discussed in the 1960s from practical viewpoints.

Karbowiak showed that, e.g., when the inner diameter of the pipe is 2 mm and $\lambda = 1$ μm, theoretical transmission loss of the lowest mode (HE_{11} mode) is 1.8 dB/km [20]. Apparently this value seemed satisfactory; however, it was also shown theoretically that even a very slight bend (with a radius of curvature of 10 km) would double the loss. At present, nobody believes that this type of waveguide is practical.

G. Dielectric Surface Waveguide

It is also possible to modify the hollow dielectric waveguide (Fig. 1.10a) to a solid type (Fig. 1.10b) so as to guide the light ray by using the total reflection at the dielectric–air boundary. This is a typical surface-waveguiding structure (see Section 2.4.4). Such a light-guiding mechanism was discovered by John Tyndall around 1870, and later investigated theoretically by Hondros and Debye in 1910 [21].

In the 1950s, the light guide as shown in Fig. 1.10b became familiar to researchers because many such fibers were bundled and used in the so-called "fiberscope" for medical uses. The individual light-guiding fiber was called the optical fiber. However, until the middle 1960s few people thought that the optical fiber could be used in long-distance optical communications, probably because of two reasons.

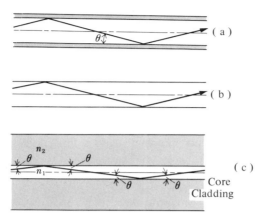

Fig. 1.10 Various axially symmetric dielectric optical waveguides: (a) hollow dielectric waveguide, (b) dielectric rod waveguide, (c) "weakly guiding" optical fiber.

First, as described in Section D, until the middle 1960s the dielectric loss of glass (tan δ) was greater than 10^{-8}, which meant a transmission loss of 500 dB/km. It was believed that this value would not be improved dramatically. Consequently, it was thought that a low-loss optical waveguide might be obtained if most of the light energy propagates outside the dielectric fiber merely "guided" by it, but could never be obtained if most of the energy propagates in the lossy fiber.

Second, it was known that in the structure shown in Fig. 1.10b, which is now called an "unclad" fiber, transmitted light pulses were much wider at the receiving end, and hence such a structure could never be a practical communication channel even if the loss were reduced. As discussed in later sections, optical fibers now being investigated for communication purposes have essentially different structures such as shown in Fig. 1.10c; in this structure the light-guiding central portion (called "core") is covered by a dielectric layer (called "cladding") having a refractive index slightly (0.1–1%) lower than that of the core. For communication purposes such a structure is more favorable than the unclad structure because the higher mode energy is easily radiated, resulting in less multimode dispersion and hence less pulse spread at the receiving end. However, until the middle 1960s, people had been unaware of the advantage of such a structure.

It was after the middle 1960s that the optical characteristics of such a "core–cladding" structure were discussed from the viewpoint of the pulse transmission; in 1971 Gloge [22] named such a structure the "weakly guiding fiber." The concept of the weakly guiding fiber was really essential in achieving dispersion small enough for practical communication purposes.

The optical fiber for communications, the subject of this book, is thus a special kind of dielectric surface waveguide, i.e., the "weakly guiding" dielectric waveguide.

1.2.4   Fourth Period—Advent of Practical Optical Fibers

The fourth period of optical communications research is characterized by the fact that researchers had a definite technical target: optical fiber communications. It is difficult to tell exactly when this period began; however, if we define its beginning by the establishment of the consensus among researchers of the superiority of the optical fiber above all other optical waveguides, the fourth period may be said to have begun in 1970 when Kapron *et al.* [23] (Corning Glass Works) reported a fiber with a 20-dB/km loss. However, before Kapron's report, there had been vigorous efforts toward the optical fiber for communication uses.

As described earlier, researchers had long been familiar with the fiber-scope [24], but until the middle 1960s very few thought that the optical fiber could be used in long-distance communications. In 1964, Nishizawa and Sasaki [25] (Tohoku University) proposed an optical fiber with a gas-lens-like focusing refractive-index distribution and filed a patent. On the other hand, Kao and Hockham [26] of Standard Telecommunication Laboratories (STL) investigated the optical loss mechanism in glass and predicted that future technological improvement would result in optical fibers with loss low enough for communications uses. They speculated that if the content of transition metal ions which played a principal role in the loss mechanism was reduced to 10^{-6}, then transmission loss due to absorption would be reduced below 20 dB/km. They also asserted that the loss would be further reduced below several dB/km if the purity of the material was further improved, making scattering loss the limiting factor. These two works were forerunners of the fiber age in the 1970s.

Nishizawa and Sasaki's proposal was followed by the theoretical work of Nishizawa and Kawakami [27] and the first successful fabrication of light-focusing glass rods and fibers by Uchida *et al.* of Nippon Electric Company (NEC) and Kitano *et al.* (Nippon Sheet Glass) in 1969 [28]. Uchida *et al.* named their self-focusing fiber the SELFOC. Shortly afterward, Pearson *et al.* [29] of Bell Laboratories (BL) reported success in preparing light-focusing glass rods. On the other hand, Kapron *et al.* [23] of Corning Glass Works (CGW) continued efforts to pursue experimentally the loss limit predicted by Kao and Hockham, and obtained a transmission loss of 20 dB/km with silica fiber produced by the outside vapor-phase oxidation (OVPO) method, which was developed for this purpose using

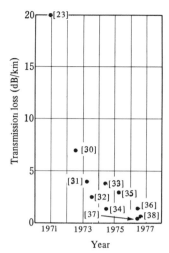

Fig. 1.11   Improvement in the transmission loss of optical fibers during 1971–1976. Numerals denote references.

the chemical vapor deposition (CVD) technique. This success triggered explosive research activities in the following years toward the development of practical optical fibers.

Research and development after 1970 is probably too familiar to us to be called history. The loss of silica fibers continued to be reduced by the improvement of the CVD method and reached 7 dB/km in 1972 (Corning Glass Works) [30], and 2.5 dB/km in 1973 (BL) [32]. The method developed by BL was named the modified chemical vapor deposition (MCVD) method (see Section 2.11.3), and became a standard method in the following years. In 1974–1975, several organizations in the United States and Japan reported the achievement of 1.5 to 2.0 dB/km. Further, at the time this book is being written, Ibaraki Electrical Communication Laboratory (ECL) and Fujikura Cable Works have reported the lowest achieved loss of 0.47 dB/km at a wavelength of 1.2 μm [37]. Typical achievements during 1971–1976 including those already described are summarized in Fig. 1.11.

On the other hand, optical fibers made of so-called compound glass also showed progress, though slower than that of silica fibers. Since the development of SELFOC by NEC in 1969, other laboratories, such as BL (U.S.A.), Ibaraki ECL (Japan), the British Post Office (U.K.), and Janaer Glaswerk Schott und Genossenschaft (JGS) (F.R.G.), have reported the improvement of soda-glass fibers until 1974. In 1975 NEC achieved a 10-dB/km loss with a soda-glass fiber having a graded-index profile (see Fig. 5.1). The loss is still much higher than that of silica fibers, and research and development are less active than that on silica fibers. However, the inherent mass-producibility of the compound-glass fibers is still attracting investigators.

1.3 Summary of the History

In the 1950s researchers believed that the push toward shorter wavelengths would finally reach the optical region. However, before 1970, nobody could tell what kind of optical waveguide would actually be used; various candidates were investigated in parallel. In the early 1970s, optical fibers having a loss below 20 dB/km appeared and entirely changed the situation. The superiority of the optical fiber above all other optical waveguides became evident. Furthermore, people realized that in the distant future the optical fiber communication system would become the most economical one among other communication systems such as the coaxial cable, microwave radio link, millimeter waveguide, and satellite. The shortage of copper resources will also accelerate the widespread use of the optical fiber communications.

1.4 Purpose and Organization of This Book

In view of the present state of the art, this book deals with, among various optical waveguides, the optical fiber for communications use. Although there are many versions of the optical fiber, only those having axially symmetrical structures (refractive-index distributions) are discussed.

The optical and electromagnetic wave aspects of optical fibers are emphasized. Materials, fabrication technologies, applications, and communication-system considerations are described only occasionally. This is because optical and wave aspects can be described in a definite manner, whereas materials, fabrication, applications, and systems are much more progress-dependent and do not yet permit textbook-like descriptions.

The basic concepts and equations required for the comprehension of the following discussions are presented in Chapter 2. In the following three chapters, the principal approaches to optical fiber analysis are discussed; the ray theory, the wave theory of uniform-core fibers, and the wave theory of nonuniform-core fibers are described in Chapters 3, 4, and 5, respectively. Attention is given to the uniform use of symbols in various analyses to assist the understanding of the analyses: ray theory, mode theory, WKB method, power-series method, variational methods, and staircase approximation method. Various analytical approaches are reviewed and compared in Chapter 6. Thus, Chapters 3–6 constitute the core of this book.

In Chapter 7, the optimum design of the refractive-index distributions of single-mode and multimode fibers are discussed. Scattering and coupling phenomena in axially nonuniform optical fibers are described in Chapter 8.

The measurements of refractive-index distribution and transmission characteristics are described in Chapters 9 and 10, respectively. Concluding remarks are given in Chapter 11.

References

1. T. Yamasaki and T. Kimoto, "History of Electrical Engineering (in Japanese)," p. 33. Ohm Publ., Tokyo, 1976. Readers outside Japan may refer, e.g., to Von Volker Aschoff, Optische Nachrichtenubertragung in Klassischen Altertum, *NTZ* **30**, No. 1, 23–28 (1977).
2. D. Gloge, Optical waveguide transmission, *Proc. IEEE* **58**, No. 10, 1513–1522 (1970).
3. G. Goubau and J. R. Christian, A new waveguide for millimeter waves, *Proc. Army Sci. Conf. U.S. Military Acad.*, West Point, New York 291–303 (1959).
4. J. W. Mink, Experimental investigations with an iris beam waveguide, *IEEE Trans. Microwave Theory Tech.* **MTT-17**, No. 1, 48–49 (1969).
5. G. Goubau and F. Schwering, On the guided propagation of electromagnetic wave beams, *IRE Trans. Antenna and Propagation* **AP-9**, No. 5, 248–256 (1961).
6. D. Gloge and W. H. Steier, Pulse shuttling in a half-mile optical lens guide, *Bell Syst. Tech. J.* **47**, 767–782 (1968).
7. D. W. Berreman, A lens or light guide using convectively distorted thermal gradient in gases, *Bell Syst. Tech. J.* **43**, 1469–1475 (1964).
8. D. Marcuse and S. E. Miller, Analysis of a tubular gas lens, *Bell Syst. Tech. J.* **43**, 1759–1782 (1964).
9. D. Gloge, Deformation of gas lenses by gravity, *Bell Syst. Tech. J.* **46**, 357–365 (1967).
10. P. Kaiser, An improved thermal gas lens for optical beam waveguides, *Bell Syst. Tech. J.* **49**, 137–153 (1970).
11. P. Kaiser, Measured beam deformations in a guide made of tubular gas lenses, *Bell Syst. Tech. J.* **47**, 179–194 (1968).
12. Y. Suematsu, Light beam waveguide using lens-like media with periodical hyperbolic temperature distribution (in Japanese), *J. IECE Jpn.* **49**, No. 3, 463–469 (1966).
13. Y. Suematsu, K. Iga, and S. Ito, A light beam waveguide using hyperbolic-type gas lenses, *IEEE Trans. Microwave Theory Tech.* **MTT-14**, No. 12, 657–665 (1966).
14. A. E. Karbowiak, New type of waveguide for light and infrared waves, *Electron. Lett.* **1**, 47–48 (1965).
15. N. Kumagai, S. Kurazono, S. Wawa, and N. Yoshikawa, Surface waveguide consisting of inhomogeneous dielectric thin-film (in Japanese), *Trans. IECE Jpn.* **51-B**, No. 3, 82–87 (1969).
16. S. Sawa and N. Kumagai, Surface wave along a circular H-bend of an inhomogeneous dielectric thin film (in Japanese), *Trans. IECE Jpn.* **52-B**, No. 3, 115–121 (1969).
17. A. E. Karbowiak, Guided wave propagation in submillimetric region, *Proc. IRE* **41**, No. 10, 1706–1711 (1958).
18. E. A. J. Marcatili and R. A. Schmeltzer, Hollow metalic and dielectric wave-guides for long-distance optical transmission and lasers, *Bell Syst. Tech. J.* **43**, No. 7, 1783–1809 (1964).
19. M. Prochazka, J. Packman, and J. Musik, Experimental investigation of a pipeline for optical communication, *Electron. Lett.* **3**, 73–74 (1967).
20. A. E. Karbowiak, *Microwaves* 37 (July 1966).
21. D. Hondros and P. Debye, Elektromagnetische Wellen an dielektrischen Drähten, *Ann. Phys. Vierte Folge* Band 32, Heft 8, pp. 465–476 (1910).

22. D. Gloge, Weakly guiding fibers, *Appl. Opt.* **10**, No. 10, 2252–2288 (1971).
23. F. P. Kapron *et al.*, Radiation loss in glass optical waveguide, *Appl. Phys. Lett.* **17**, 423–425 (1970).
24. N. S. Kapany, "Optical Fibers—Principles and Applications." Academic Press, New York, 1967.
25. J. Nishizawa *et al.*, Optical waveguiding structure containing materials of relatively high refractive index and relatively low refractive index, Japanese Patent filed in 1964, No. S39-64040.
26. K. C. Kao and G. A. Hockham, Dielectric-fiber surface waveguides for optical frequencies, *Proc. IEE* (*London*) **113**, 1151–1158 (1966).
27. S. Kawakami and J. Nishizawa, An optical waveguide with the optimum distribution of the refractive index with reference to waveform distortion, *IEEE Trans. Microwave Theory Tech.* **MTT-16**, No. 10, 814–818 (1968).
28. T. Uchida, M. Furukawa, I. Kitano, K. Koizumi, and H. Matsumura, A light-focusing fiber guide, *IEEE J. Quantum Electron.* **QE-5**, 331 (1969).
29. A. D. Pearson, W. G. French, and E. G. Rawson, Preparation of a light focusing glass rod by ion-exchange techniques, *Appl. Phys. Lett.*, **15**, No. 2, 76–77 (1969).
30. D. B. Keck, P. C. Schultz, and F. Zimar, Attenuation of multimode glass optical waveguides, *Appl. Phys. Lett.* **21**, No. 5. 215–217 (1972).
31. D. B. Keck, R. D. Maurer, and P. C. Schultz, On the ultimate lower limit of attenuation in glass optical waveguides, *Appl. Phys. Lett.* **22**, No. 7, 307–309 (1973).
32. P. Kaiser, Spectral losses of unclad fibers made from high-grade vitreous silica, *Appl. Phys. Lett.* **23**, No. 1, 45–46 (1973).
33. D. W. Black, J. Irven *et al.*, Measurements on waveguide properties of GeO_2–SiO_2-core optical fibers, *Electron. Lett.* **10**, No. 12, 239–240 (1974).
34. W. G. French, J. B. MacChesney *et al.*, Optical waveguides with very low losses, *Bell Syst. Tech. J.* **53**, No. 5, 951–954 (1974).
35. D. N. Payne and W. A. Glambling, A borosilicate-cladded phosphosilicate-core optical fiber, *Opt. Commun.* **13**, No. 4, 422–425 (1975).
36. P. Geittner, D. Krüppers, and H. Lydtin, Low-loss optical fibers prepared by plasma-activated chemical vapor deposition (CVD), *Appl. Phys. Lett.* **28**, No. 11, 645–646 (1976).
37. M. Horiguchi, Spectral losses of low-OH-content optical fibers, *Electron. Lett.* **12**, No. 12, 310–311 (1976).
38. K. Okamura *et al.*, Compensation technique of index depression of optical fibers (in Japanese), Paper of Technical Group, IECE Japan, No. OQE76-35 (1976).
39. "From Semaphore to Satellite," International Telecommunication Union, Geneva, 1965.

2 | Basic Concepts and Equations

Basic concepts and equations required for the comprehension of the following chapters are presented. Loss mechanisms in optical fibers and fabrication technologies are rather briefly described.

2.1 Introduction

The basic concepts and equations of electromagnetic wave theory and optics required for the comprehension of the following chapters are presented. These are Maxwell's equations, the wave equation, various solutions of the wave equation, the concept of surface waves, phase velocity, group velocity, the complex Poynting vector, the eikonal equation, the ray equation, and reflection and refraction at boundaries between two media.

The material aspects of optical fibers required for the comprehension of their wave aspects are then described very briefly; these are the absorption-loss mechanism in optical fibers and the fabrication technologies.

2.2 Wave Equation

2.2.1 Maxwell's Equations

Maxwell's equations are written in terms of the electric field $\mathbf{E}$, magnetic field $\mathbf{H}$, electric flux density $\mathbf{D}$, and magnetic flux density $\mathbf{B}$ as [1]

$$\nabla \times \mathbf{E} = -\partial \mathbf{B}/\partial t, \tag{2.1}$$

$$\nabla \times \mathbf{H} = \partial \mathbf{D}/\partial t. \tag{2.2}$$

These four vectors are also related by the equations

$$\mathbf{D} = \varepsilon\mathbf{E}, \tag{2.3}$$

$$\mathbf{B} = \mu\mathbf{H}, \tag{2.4}$$

where ε and μ denote the permittivity and permeability of the medium, respectively.

If the vectors $\mathbf{E}$, $\mathbf{H}$, $\mathbf{D}$, and $\mathbf{B}$ are sinusoidal functions of time, they are usually represented by the complex amplitudes, i.e., the so-called phasors. For example, when $\mathbf{E}(t)$ varies sinusoidally with an angular frequency ω, we express $\mathbf{E}(t)$ as

$$\mathbf{E}(t) = \text{Re}[\mathbf{E}e^{j\omega t}], \tag{2.5}$$

and let the complex amplitude $\mathbf{E}$ at $t = 0$ represent the time-dependent variable $\mathbf{E}(t)$. The complex amplitude $\mathbf{E}$ is often called a phasor.

2.2.2 Derivation of Wave Equation

If we consider vectors $\mathbf{E}$, $\mathbf{H}$, $\mathbf{D}$, and $\mathbf{B}$ to be phasors, the partial differentiation with respect to time $(\partial/\partial t)$ can be replaced by $j\omega$. Therefore, we may write

$$\nabla \times \mathbf{E} = -j\omega\mu\mathbf{H}, \tag{2.6}$$

$$\nabla \times \mathbf{H} = j\omega\varepsilon\mathbf{E}. \tag{2.7}$$

If we further assume that ε and μ are not functions of space but constants, we obtain the following equations for $\mathbf{E}$ and $\mathbf{H}$:

$$\nabla \times \nabla \times \mathbf{E} = -j\omega\mu\nabla \times \mathbf{H} = \omega^2\varepsilon\mu\mathbf{E}, \tag{2.8}$$

$$\nabla \times \nabla \times \mathbf{H} = j\omega\varepsilon\nabla \times \mathbf{E} = \omega^2\varepsilon\mu\mathbf{H}. \tag{2.9}$$

Because these two equations for $\mathbf{E}$ and $\mathbf{H}$ are of the same form, we henceforth consider only $\mathbf{E}$. By using a vector identity,

$$\nabla \times \nabla \times \mathbf{A} = \nabla(\nabla \cdot \mathbf{A}) - \nabla^2\mathbf{A}, \tag{2.10}$$

and noting that Eq. (2.7) leads directly to

$$\nabla \cdot \mathbf{E} = 0, \tag{2.11}$$

we may rewrite Eq. (2.8) as

$$\nabla^2\mathbf{E} + k^2\mathbf{E} = 0, \tag{2.12}$$

where

$$k = \omega\sqrt{\varepsilon\mu}. \tag{2.13}$$

Equation (2.12) and the parameter k are called the wave equation and the wave number of the space, respectively. As shown later, k is equal to the phase constant of a plane wave propagating in that space.

Equations identical to Eq. (2.12) hold also for **H**, **D**, and **B**. If we further represent any one Cartesian component of the four vectorial variables (E_x, for example) by V, we may write

$$\nabla^2 V + k^2 V = \frac{\partial^2 V}{\partial x^2} + \frac{\partial^2 V}{\partial y^2} + \frac{\partial^2 V}{\partial z^2} + k^2 V = 0. \tag{2.14}$$

Note that V is a scalar phasor.

2.3   Solution of Wave Equation

To solve Eq. (2.14), we first write

$$V = V_0 e^{-\gamma_x x} e^{-\gamma_y y} e^{-\gamma_z z}, \tag{2.15}$$

and substitute this into Eq. (2.14) to obtain

$$\gamma_x^2 + \gamma_y^2 + \gamma_z^2 + k^2 = 0. \tag{2.16}$$

This is an equation describing the propagation in a homogeneous medium, and is called the characteristic equation of the wave. Only waves having γ_x, γ_y, and γ_z satisfying this relation may exist in the medium. Parameters γ_x, γ_y, and γ_z are called the propagation constants with respect to the x, y, z directions, respectively. Generally the γ are complex parameters and are expressed as

$$\gamma_i = j\beta_i + \alpha_i \qquad (i = x, y, z), \tag{2.17}$$

where β_i and α_i are called the phase constant and attenuation constant with respect to the three directions, respectively. (Customarily, β_i is often called the propagation constant.) When the suffixes are omitted, α and β usually denote the constants with respect to the direction of the waveguide axis, which in the following discussions is taken in the z direction unless otherwise stated.

2.4   Classification of Wave Equation Solutions

Electromagnetic waves of technical interest are classified into four categories with respect to the values of the transverse propagation constants (γ_x and γ_y): (1) plane waves, (2) waves propagating between two conductors,

(3) waves propagating in a hollow conducting waveguide, and (4) surface waves [2]. Waves propagating in optical fibers are surface waves.

2.4.1 Plane Waves

When the wave is uniform in the transverse (x and y) directions, $\partial V/\partial x = \partial V/\partial y = 0$, and hence

$$\gamma_x = \gamma_y = 0. \tag{2.18}$$

Therefore,

$$\beta_z = k \equiv \omega\sqrt{\varepsilon\mu}. \tag{2.19}$$

This means that the phase constant of a plane wave in a homogeneous medium is equal to the wave number of that medium.

The parameter k is called the wave number because it is equal to 2π times the number of waves in a unit length; i.e., $k = 2\pi/\lambda$, where λ denotes the wavelength. To derive this relation, we note that if we write

$$V(t) = \text{Re}[V_0 e^{-j\beta_z z} e^{j\omega t}], \tag{2.20}$$

the phase velocity v_p is derived from the constant phase relation

$$-j\beta_z z + j\omega t = \text{const} \tag{2.21}$$

as

$$v_p = dz/dt = \omega/\beta = 1/\sqrt{\varepsilon\mu}. \tag{2.22}$$

This equation directly leads to

$$\lambda = v_p/f = 2\pi/\omega\sqrt{\varepsilon\mu} = 2\pi/k, \tag{2.23}$$

where f is the frequency. Equation (2.23) is the relation to be proved.

In the vacuum, $\varepsilon = \varepsilon_0 = 8.854 \times 10^{-12}$ [F/m], $\mu = \mu_0 = 4\pi \times 10^{-7}$ [H/m], and $v_p = 1/\sqrt{\varepsilon_0\mu_0} = c$ (light velocity) $= 2.9998 \times 10^8$ [m/s].

2.4.2 Waves Propagating between Two Conductors

Waves with $E_z = H_z = 0$ are called TEM (transverse electric and magnetic) waves. A plane wave is always a TEM wave. For TEM waves, the transverse electric and magnetic fields E_t, H_t are expressed as gradients of scaler functions of x and y as [3]

$$\mathbf{E}_t = -V(z)\nabla\phi(x, y), \tag{2.24}$$

$$\mathbf{H}_t = -V(z)\nabla\psi(x, y). \tag{2.25}$$

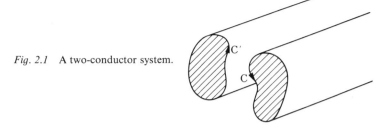

Fig. 2.1   A two-conductor system.

On the other hand, $\nabla \cdot \mathbf{E} = \nabla \cdot \mathbf{H} = 0$. Hence, using a vector identity $\nabla \cdot (\nabla \phi) = \nabla^2 \phi$, we can prove that ϕ and ψ satisfy two-dimensional Laplace equations,

$$\nabla^2 \phi = 0, \qquad \nabla^2 \psi = 0. \tag{2.26}$$

It can be shown that a TEM wave can exist only in a space between two conductors, such as two parallel conductors (see Fig. 2.1) or coaxial cylinders. (The plane wave is also a TEM wave; this is a rather special case in which we may consider that the wave propagates between two ficticious plates existing at both ends of electric lines of force.) The scalar variables $\phi(x, y)$ and $\psi(x, y)$ correspond to the electric and magnetic scalar potentials, respectively. When only one conductor is present, ϕ is zero everywhere and the TEM wave can never exist.

For a TEM wave, if we write

$$\phi = \phi_0 e^{-\gamma_x x} e^{-\gamma_y y}, \tag{2.27}$$

then from Eq. (2.26), we obtain

$$\gamma_x^2 + \gamma_y^2 = 0, \tag{2.28}$$

which leads to $\beta_z = k = \omega/c$, showing that the phase velocity is equal to that of the plane wave.

2.4.3.   Waves Propagating in a Hollow Conducting Waveguide

Consider cases when γ_x and γ_y are both imaginery, or when one of these is imaginary and the other is zero. Such conditions hold for waves propagating in a hollow conducting waveguide because the fields in such a structure are harmonic at least in one transverse direction. The best-known example is the TE_{10} -mode in a rectangular waveguide as shown in Fig. 2.2, for which $\gamma_x = j\pi/a$ and $\gamma_y = 0$.

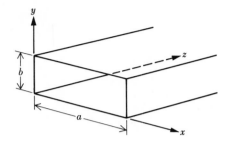

Fig. 2.2   A hollow conducting waveguide.

In such cases, from Eq. (2.16),

$$-\gamma_z^2 < k^2. \tag{2.29}$$

This leads to $\beta_z < k$ if $\alpha_z = 0$ is assumed, and hence to

$$v_{\rm p} > 1/\sqrt{\varepsilon\mu}, \tag{2.30}$$

which shows that the phase velocity of the wave propagating in a hollow conducting waveguide is always faster than the plane wave propagating in the same medium filling the waveguide.

It is also known that such waves are classified into

(1)   TE (transverse electric) waves, for which $E_z = 0$ but $H_z \neq 0$, and
(2)   TM (transverse magnetic) waves, for which $H_z = 0$ but $E_z \neq 0$.

2.4.4   Surface Waves

Finally, consider cases when the electromagnetic field attenuates exponentially in the transverse directions in the vicinity of a waveguiding structure, as shown in Fig. 2.3. Such waves are called generically surface waves, regardless the physical construction of the waveguiding structure. All waves propagating in an optical fiber are surface waves.

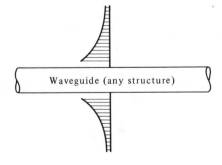

Waveguide (any structure)

Fig. 2.3   Concept of the surface wave.

The exponential transverse attenuation implies that γ_x and γ_y are real. Therefore, for a surface wave

$$-\gamma_z^2 > k^2, \tag{2.31}$$

which, if $\alpha_z = 0$ is assumed, leads to

$$v_p < 1/\sqrt{\varepsilon\mu}. \tag{2.32}$$

Thus, we may conclude that a surface wave always has a phase velocity slower than that of the plane wave propagating in the same medium used in the outside "exponentially attenuating" region of the waveguide.

It is also known that in surface-wave structures, in addition to the TE and TM modes, modes having both E_z and H_z can propagate. These are called hybrid modes. Hybrid modes are further classified into HE modes and EH modes as described in Chapter 4. Usually, analyses of surface waveguides are much more complicated than those of hollow conducting waveguides.

2.5 Phase Velocity and Group Velocity

When the phase velocity v_p of a wave is given as a function of the angular frequency ω of the transmitted signal, the group velocity v_g is given as

$$v_g = (d\beta/d\omega)^{-1}, \tag{2.33}$$

where $\beta = \omega/v_p$. The proof of this equation follows.

When loss is absent, the group velocity is given as the velocity of the envelope of the wave when it is amplitude modulated. We therefore consider a sinusoidally modulated signal at $z = 0$,

$$f(0, t) = A(1 + m \sin pt)\sin \omega t, \tag{2.34}$$

and investigate how it propagates. This equation may be rewritten as

$$f(0, t) = A \sin \omega t + \tfrac{1}{2}Am\{\cos(\omega - p)t - \cos(\omega + p)t\}, \tag{2.35}$$

which means that the wave consists of three frequency components: ω, $(\omega - p)$, and $(\omega + p)$. Therefore, if the phase constant is β for ω, $(\beta - \delta)$ for $(\omega - p)$, and $(\beta + \delta)$ for $(\omega + p)$, where $p \ll \omega$ and $\delta \ll \beta$, then the modulated wave will become, at a position z,

$$\begin{aligned}
f(z, t) &= A \sin(\omega t - \beta z) + \tfrac{1}{2}Am[\cos\{(\omega - p)t - (\beta - \delta)z\}, \\
&\quad -\cos\{(\omega + p)t - (\beta + \delta)z\}] \\
&= A\{1 + m \sin(pt - \delta z)\} \sin(\omega t - \beta z). \tag{2.36}
\end{aligned}$$

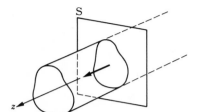

Fig. 2.4   Transmission power flowing in the z direction.

The velocity of the envelope is given, from the condition $pt - \delta z = \text{const}$, as

$$v_g = \lim_{\delta \to 0} (p/\delta) = (d\beta/d\omega)^{-1}, \tag{2.37}$$

which is the relation to be proved.

2.6   Propagating Power

When an electromagnetic wave propagates along a waveguiding structure as shown in Fig. 2.4, the electromagnetic power carried by this wave is expressed as

$$P = \int_{-\infty}^{+\infty} \int_{-\infty}^{+\infty} \text{Re}(S_z)\, dxdy, \tag{2.38}$$

where S_z is the z component of a vector defined in terms of phasors representing electric and magnetic fields,

$$\mathbf{S} = \tfrac{1}{2}(\mathbf{E} \times \mathbf{H}^*), \tag{2.39}$$

the asterisk (*) denoting the complex conjugate. The vector $\mathbf{S}$ is called the complex Poynting vector.

To prove Eq. (2.38), we take for granted that the power-flow density at an instant is given by the instantaneous Poynting vector

$$\mathbf{S}(t) = \mathbf{E}(t) \times \mathbf{H}(t). \tag{2.40}$$

In this equation $\mathbf{E}(t)$ and $\mathbf{H}(t)$ denote instantaneous fields as functions of time t. The average power-flow density in an alternating field is then given, by denoting a time average by $\langle \ \rangle$ and phasor representations of fields by $\mathbf{E}$ and $\mathbf{H}$, as

$$
\begin{aligned}
\langle \mathbf{S}(t) \rangle &= \langle \mathbf{E}(t) \times \mathbf{H}(t) \rangle = \langle \text{Re}(\mathbf{E}e^{j\omega t}) \times \text{Re}(\mathbf{H}e^{j\omega t}) \rangle \\
&= \tfrac{1}{4} \langle (\mathbf{E}e^{j\omega t} + \mathbf{E}^* e^{-j\omega t}) \times (\mathbf{H}e^{j\omega t} + \mathbf{H}^* e^{-j\omega t}) \rangle \\
&= \tfrac{1}{2} \text{Re}(\mathbf{E} \times \mathbf{H}^*).
\end{aligned}
\tag{2.41}
$$

Using Eqs. (2.39) and (2.41), we may derive Eqs. (2.38).

2.7 Eikonal Equation

Most of the problems dealt with in this book are related to the behavior of light beams in a spatially inhomogeneous medium, in which the refractive index is expressed generally as $n(x, y, z)$. Two approaches are known for such problems. The first is to investigate the path of a ray of light traveling through the medium. This approach is called geometrical optics. The second is to solve Maxwell's equations or the wave equation to obtain solutions expressing the wave behavior. The first approach is discussed in Chapter 3, and the second is described in Chapters 4, 5, 7, and 8. The relation between the two approaches is discussed in Chapter 6.

The ray concept is applicable only to those cases in which the beam diameter is much larger than the wavelength of light. A beam having a diameter comparable to the wavelength can never be treated as a ray, because such a beam is subject to an immediate divergence due to diffraction, which is a phenomenon beyond geometrical optics. On the contrary, if the beam diameter is much larger than the wavelength of light, an approximate wave equation expressing the ray path can be derived from Maxwell's equations. The real scalar variable used in such an equation to express the wavefront phase is called the "eikonal," and the equation is called the "eikonal equation." The discussions in Chapter 3 are essentially based on this equation.

The derivation of the eikonal equation follows [4]. We first consider that a uniform plane wave travels in the direction $\mathbf{s}$ in a homogeneous space having a refractive index n. Then the phasors representing the electric and magnetic fields are expressed, in terms of the position vector $\mathbf{r}$, as

$$\mathbf{E} = \mathbf{e}e^{-jk_0 n(\mathbf{s} \cdot \mathbf{r})}, \tag{2.42}$$

$$\mathbf{H} = \mathbf{h}e^{-jk_0 n(\mathbf{s} \cdot \mathbf{r})}, \tag{2.43}$$

where $\mathbf{e}$ and $\mathbf{h}$ denote $\mathbf{E}$ and $\mathbf{H}$ at $\mathbf{r} = 0$ (origin), respectively, k_0 is the vacuum wavenumber, and $(\mathbf{s} \cdot \mathbf{r})$ expresses the scalar product.

When n is no longer constant but is given as a function of position $\mathbf{r}$, $\mathbf{E}$ and $\mathbf{H}$ can be expressed in forms similar to Eqs. (2.42) and (2.43) as

$$\mathbf{E} = \mathbf{e}(\mathbf{r})e^{-jk_0 \phi(\mathbf{r})}, \tag{2.44}$$

$$\mathbf{H} = \mathbf{h}(\mathbf{r})e^{-jk_0 \phi(\mathbf{r})}. \tag{2.45}$$

In these equations the function $\phi(\mathbf{r})$ is a scalar real function expressing the phase at position $\mathbf{r}$; this function is called the "eikonal" or sometimes the "optical path." Functions $\mathbf{e}(\mathbf{r})$ and $\mathbf{h}(\mathbf{r})$ are vectorial real functions of position $\mathbf{r}$. (If we also wish to express circularly or elliptically polarized waves in a unified manner, we may consider that each $\mathbf{e}$ and $\mathbf{h}$ consists of two perpendicular components: a real vector and an imaginary vector. However, since a circularly or elliptically polarized wave can be expressed by two linearly polarized waves having different phases, we consider only linearly polarized waves.)

We first show that when k is very large, i.e., when the space under consideration is much greater than the wavelength, we may obtain a differential equation only for ϕ and independent of **e** and **h**. From Eq. (2.44), we obtain

$$\nabla \times \mathbf{E} = (\nabla \times \mathbf{e} - jk_0 \nabla\phi \times \mathbf{e})e^{-jk_0\phi}. \tag{2.46}$$

Similarly, for the magnetic field, we obtain from Eq. (2.45)

$$\nabla \times \mathbf{H} = (\nabla \times \mathbf{h} - jk_0 \nabla\phi \times \mathbf{h})e^{-jk_0\phi}. \tag{2.47}$$

On the other hand, putting Eqs. (2.44) and (2.45) into Eqs. (2.6) and (2.7) leads to

$$\nabla \times \mathbf{E} = -jk_0 c\mu\mathbf{H} = -jk_0 c\mu\mathbf{h}e^{-jk_0\phi}, \tag{2.48}$$

$$\nabla \times \mathbf{H} = jk_0 c\varepsilon\mathbf{E} = jk_0 c\varepsilon\mathbf{e}e^{-jk_0\phi}, \tag{2.49}$$

where c is the velocity of light. Combining Eq. (2.46) with (2.48), and (2.47) with (2.49), we obtain

$$\nabla\phi \times \mathbf{e} - c\mu\mathbf{h} = (1/jk_0)\nabla \times \mathbf{e}, \tag{2.50}$$

$$\nabla\phi \times \mathbf{h} + c\varepsilon\mathbf{e} = (1/jk_0)\nabla \times \mathbf{h}. \tag{2.51}$$

We now assume that the space under consideration and the lengths associated with the behavior of the ray (e.g., the beam diameter and radius of curvature of the beam axis) are all much greater than the wavelength. In such a case we may consider that the wavelength is extremely short, or the frequency is extremely high, or in other words k_0 ($= \omega/c$) tends to infinity. Hence, the right-hand sides of Eqs. (2.50) and (2.51) are both zero, so that

$$\nabla\phi \times \mathbf{e} - c\mu\mathbf{h} = \mathbf{0}, \tag{2.52}$$

$$\nabla\phi \times \mathbf{h} + c\varepsilon\mathbf{e} = \mathbf{0}. \tag{2.53}$$

These equations given relations between the directions of $\nabla\phi$, **h**, and **e**; evidently, each of them is normal to the other two. Hence,

$$\mathbf{e} \cdot \nabla\phi = 0, \tag{2.54}$$

$$\mathbf{h} \cdot \nabla\phi = 0. \tag{2.55}$$

Equations (2.52)–(2.55) are the starting equations for the following discussions. Eliminating **h** from Eqs. (2.52) and (2.53) and using vector identities, we obtain

$$(1/c\mu)[(\mathbf{e} \cdot \nabla\phi)\nabla\phi - \mathbf{e}(\nabla\phi)^2] + c\varepsilon\mathbf{e} = 0, \tag{2.56}$$

which is simplified further by using Eq. (2.54) to

$$(\nabla\phi)^2 = c^2\varepsilon\mu = n^2. \tag{2.57}$$

In Cartesian coordinates, therefore,

$$(\partial\phi/\partial x)^2 + (\partial\phi/\partial y)^2 + (\partial\phi/\partial z)^2 = n^2(x, y, z). \tag{2.58}$$

This is the equation that governs the eikonal ϕ. As shown in the next section, an equieikonal surface on which

$$\phi(\mathbf{r}) = \text{const} \tag{2.59}$$

corresponds to a wavefront of light, and the ray direction is given by a curve normally intersecting the equieikonal surfaces. Equation (2.57) is called the "eikonal equation"; this is the basic equation of geometrical optics in inhomogeneous media.

2.8 Ray Equations

From the eikonal equation we derive equations which should be called ray equations [4]. As described in Section 2.6, the power-flow density along a light beam is given as the real part of the complex Poynting vector as

$$\mathbf{S} = \text{Re}[\tfrac{1}{2}\mathbf{E} \times \mathbf{H}^*]. \tag{2.60}$$

This can be rewritten, by using Eqs. (2.44) and (2.45), as

$$\mathbf{S} = \text{Re}[\tfrac{1}{2}\mathbf{e} \times \mathbf{h}^*], \tag{2.61}$$

and further by using Eq. (2.52) and a vector identity as

$$\mathbf{S} = (1/2c\mu)\,\text{Re}[(\mathbf{e} \cdot \mathbf{e}^*)\nabla\phi - (\mathbf{e} \cdot \nabla\phi)\mathbf{e}^*]. \tag{2.62}$$

However, since the second term within brackets is zero [see Eq. (2.54)],

$$\mathbf{S} = (1/2c\mu)\,\text{Re}[(\mathbf{e} \cdot \mathbf{e}^*)\nabla\phi] = (1/2c\mu)(\mathbf{e} \cdot \mathbf{e}^*)\nabla\phi. \tag{2.63}$$

Since $(\mathbf{e} \cdot \mathbf{e}^*)$ in this equation is a real scalar quantity corresponding to the mean stored energy of the electric field, this equation gives the energy flow direction, i.e., the ray direction coincides with that of $\nabla\phi$. Further, from the eikonal equation (Eq. (2.57)),

$$|\nabla\phi| = n. \tag{2.64}$$

Hence, the unit vector along the ray $\mathbf{s}$, is expressed as

$$\mathbf{s} = \nabla\phi/|\nabla\phi| = \nabla\phi/n. \tag{2.65}$$

An alternative expression for the ray is obtained by using a curved coordinate s along the ray. Since

$$d\mathbf{r}/ds = \mathbf{s}, \tag{2.66}$$

Eq. (2.65) can be rewritten as

$$n(x,\, y,\, z)\, d\mathbf{r}/ds = \nabla\phi. \tag{2.67}$$

After some algebraic computations described in Appendix 2A.1, Eq. (2.67) can further be rewritten as [5]

$$(d/ds)[n(x,\, y,\, z)\mathbf{s}] = \nabla n, \tag{2.68}$$

or

$$(d/ds)[n(x,\, y,\, z)\, d\mathbf{r}/ds] = \nabla n. \tag{2.69}$$

Equations (2.68) and (2.69) are basic equations determining the ray path in an inhomogeneous medium and will be used as starting equations in Chapter 3. These equations will hereafter be called ray equations.

2.9   Refraction and Reflection at a Boundary between Two Media

2.9.1   Snell's Law

We consider a ray, as shown in Fig. 2.5, that is incident on a boundary between medium 1 (refractive index n_1) and medium 2 (refractive index n_2), at an angle θ from medium 1. Part of the light energy is reflected at an angle θ'', and the other part is refracted into medium 2 at an angle θ'. The reflection

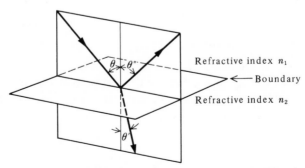

Fig. 2.5   Refraction and reflection at boundary between two dielectric media.

angle $\theta'' = \theta$, and the refraction angle θ' obeys

$$\sin \theta / \sin \theta' = n_2/n_1. \tag{2.70}$$

This relation is called Snell's law.

The principal purpose of this section is to give power relations (reflection and transmission coefficients) at such a boundary between two media. These relations as well as Snell's law can be derived from Maxwell's equations; the derivation, however, is omitted because it can be found in standard textbooks on electromagnetism [3] or optics [4]. Only the formulas and some important conclusions are presented.

2.9.2 Electric Field Normal to Incidence Plane

The power reflection coefficient R_E and power transmission coefficient T_E are [4, 5]

$$R_E = \frac{[n_1 \cos \theta - \sqrt{n_2^2 - n_1^2 \sin^2 \theta}]^2}{[n_1 \cos \theta + \sqrt{n_2^2 - n_1^2 \sin^2 \theta}]^2}, \tag{2.71}$$

$$T_E = \frac{4n_1 \cos \theta \sqrt{n_2^2 - n_1^2 \sin^2 \theta}}{[n_1 \cos \theta + \sqrt{n_2^2 - n_1^2 \sin^2 \theta}]^2}. \tag{2.72}$$

From conservation of power,

$$R_E + T_E = 1. \tag{2.73}$$

Note that the preceding relations hold only for the incident angle θ which makes the square root in Eqs. (2.71) and (2.72) real.

Consider the conditions for making the square root real. When $n_1 < n_2$, the radical is real for all θ lying within the range $0° < \theta < 90°$. When $n_2 < n_1$, the critical angle is defined as

$$\theta_c = \sin^{-1}(n_2/n_1); \tag{2.74}$$

when $\theta < \theta_c$ the radical is real, but when $\theta_c < \theta$ it is imaginary and Eqs. (2.71) and (2.72) do not hold. If we investigate in detail the field configuration in the vicinity of the boundary for the latter case, we will find that the propagation constant in medium 2 is real, i.e., the wave penetrates into medium 2 with an exponential decay. In geometrical optics such a phenomenon is known as total reflection.

When total reflection occurs, R_E and T_E are no longer expressed by Eqs. (2.71) and (2.72) but simply as

$$R_E = 1, \tag{2.75}$$

$$T_E = 0. \tag{2.76}$$

If the amplitude reflection coefficient is denoted by R_A, of course $|R_A| = 1$, but R_A is complex. This means that total reflection is accompanied by a certain amount of phase shift, which is called the Goos–Haenchen shift [6] (see Section 3.2.4).

When the wave is normally incident on the boundary ($\theta = 0$),

$$R_E = (n_1 - n_2)^2/(n_1 + n_2)^2, \tag{2.77}$$

$$T_E = 4n_1n_2/(n_1 + n_2)^2. \tag{2.78}$$

2.9.3   Electric Field Parallel to Incidence Plane

In these cases the power reflection and power transmission coefficients are given as

$$R_H = \frac{[n_2\cos\theta - (n_1/n_2)\sqrt{n_2^2 - n_1^2\sin^2\theta}]^2}{[n_2\cos\theta + (n_1/n_2)\sqrt{n_2^2 - n_1^2\sin^2\theta}]^2}, \tag{2.79}$$

$$T_H = \frac{4n_1\cos\theta\sqrt{n_2^2 - n_1^2\sin^2\theta}}{[n_2\cos\theta + (n_1/n_2)\sqrt{n_2^2 - n_1^2\sin^2\theta}]^2}. \tag{2.80}$$

Similarly, we have

$$R_H + T_H = 1, \tag{2.81}$$

and total reflection occurs when $\theta_c < \theta$, where the critical angle θ_c is given by Eq. (2.74). The magnitude of the Goos–Haenchen shift will be shown in Section 3.2.4. When $\theta = 0$, R_H and T_H agree with R_E and T_E, and are given by Eqs. (2.77) and (2.78), respectively.

A phenomenon which takes place only when the electric field is parallel to the plane of incidence is total transmission at the Brewster angle. Substituting

$$\theta = \sin^{-1}(n_2/\sqrt{n_1^2 + n_2^2}) \tag{2.82}$$

into Eq. (2.79), we obtain $R_H = 0$ and $T_H = 1$, which mean that the entire incident power is transmitted into medium 2. The angle given by Eq. (2.82) is called the Brewster angle. When the ray travels from medium 2 into medium 1, the Brewster angle is $\theta' = \sin^{-1}(n_1/\sqrt{n_1^2 + n_2^2})$; this angle and the angle given by Eq. (2.82) satisfy Snell's law.

2.10   Transmission Loss Mechanisms in Optical Fibers

Minimum prerequisite knowledge of loss mechanisms in optical fibers is presented here. Silica-glass fibers will mainly be considered.

2.10.1   Factors Generating Loss

In the wavelength region 0.8–1.6 μm which will be the most probable range used for optical communication systems in the near future, factors generating transmission loss are classified as follows [7]:

Intrinsic loss mechanisms
A. Tail of infrared absorption by Si–O coupling
B. Tail of ultraviolet absorption due to electron transition
C. Rayleigh scattering due to the spatial fluctuation of refractive index
Absorption by impurities
A. Absorption by molecular vibration of OH
B. Absorption by transition metals
Structural imperfections
A. Geometrical nonuniformity at core–cladding boundary
B. Imperfection at connection or splicing between fibers

Figure 2.6 shows the typical loss characteristics of low-loss silica fibers [8]. It is seen that, as the result of superposition of various losses already mentioned, the minimum loss (= 0.5 dB/km) is obtained at about $\lambda = 1.2$ μm. The contribution of each factor is now briefly described.

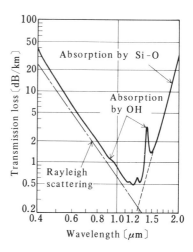

Fig. 2.6 An example of the transmission-loss characteristics of low-loss optical fibers (after Izawa [2]).

2.10.2   Intrinsic Loss Mechanisms

A. Tail of Infrared Absorption

The Si–O coupling has absorption at wavelengths of 9, 12.5, and 21 μm. The tail of these absorption spectra are shown at the right of Fig. 2.6.

B. Tail of Ultraviolet Absorption

The tail of the ultraviolet absorption having the absorption edge at 160 nm extends to the 1-μm region. This loss is estimated at about 0.3 dB/km at $\lambda = 1$ μm, by subtracting the Rayleigh scattering proportional to λ^{-4} (see Section C) from the loss shown in Fig. 2.6.

C. Rayleigh Scattering

All transparent materials have intrinsic spatial density fluctuation [9]. In the case of glass, this fluctuation is produced by the fixation of thermal density fluctuation of molten material at the instance of solidification. The density fluctuation causes the refractive index fluctuation and consequently the Rayleigh scattering. The attenuation constant due to this scattering is

$$\alpha = (8\pi^3/3\lambda^4)(n^2 - 1)^2 kT\beta, \tag{2.83}$$

where λ denotes the wavelength, n the refractive index, k the Boltzmann constant, T the solidification temperature, and β the equitemperature compressibility. This attenuation is proportional to λ^{-4} as seen in Fig. 2.6, and is typically 0.6 dB/km at $\lambda = 1$ μm.

2.10.3 Absorption by Impurities

A. Absorption by Hydroxyl Group

The fundamental molecular vibration of OH (hydroxyl group) is present at $\lambda = 2.8$ μm; hence, its second and third harmonics occur at 1.38 and 0.95 μm, respectively. The absorption spectra due to these harmonics are seen in Fig. 2.6. To suppress the second harmonic absorption below the SiO absorption, the water content must be below 10 ppb (10^{-8}) [7].

B. Absorption by Transition Metals

So far, absorption by Cr, Mn, Co, Fe, Ni, Cu, and V has been observed. Among these the effects of V and Cr usually predominate at $\lambda > 0.8$ μm. However, in recent low-loss fibers, this absorption is practically negligible because the impurity content has been satisfactorily reduced.

2.10.4 Structural Imperfections

The effects of geometrical nonuniformity at the core–cladding boundary will be discussed in Chapter 8. The effects of the imperfection at connection and splicing are not addressed in this book.

2.11   Optical Fiber Manufacture

2.11.1   Classification of Manufacturing Processes

Optical fibers may be classified into two major groups, one having a silica-glass core and the other having a compound-glass core.

The compond-glass fiber is usually manufactured by the double-crucible (or triple-crucible) method which permits high drawing speed; hence, this fiber is considered more economical than its competitor, the silica-glass fiber. However, loss is relatively high; the lowest value ever reported is 5 dB/km. Many kinds of manufacturing processes of silica-glass fibers are being investigated; among them the best established method is the modified chemical vapor deposition (MCVD) method mentioned in Section 1.2.4. The double-crucible method and the MCVD method are described.

2.11.2   Double-Crucible Method

We prepare a double crucible made of platinum as shown in Fig. 2.7. The core and cladding materials are put into the inner and outer vessels of the double crucible, respectively, and then melted and drawn to form a fiber. For example, when the core and cladding materials are doped with Ta^+ and Na^+, respectively, the refractive index in the core becomes slightly higher than that in the cladding, and thus a "weakly guiding fiber" is produced. When a graded-index fiber (see Section 5.1) is to be produced, the drawing temperature is controlled so that ion exchange takes place between Ta^+ and Na^+, causing a gradual index variation at the core–cladding boundary.

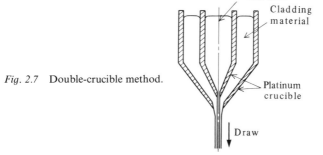

Fig. 2.7   Double-crucible method.

The double-crucible method features low cost and mass-producibility because the preprocess required in the MCVD method is not necessary; the fiber is produced at once at fairly high drawing speed (1–3 m/s).

2.11.3   MCVD Method

An example using boron as the dopant is described. As shown in Fig. 2.8, a slowly rotating silica tube is heated by a moving burner, and $SiCl_4$, O_2, and BCl_3 gases are fed into the tube. As the burner moves repeatedly along the tube, SiO_2–B_2O_3 deposits over the entire inner surface of the tube by chemical reaction of the gases. This material becomes the cladding material because it has a refractive index lower than that of pure silica (SiO_2). Next, the BCl_3 gas is stopped, and the process is resumed; the newly deposited layer consisting of pure silica becomes the core.

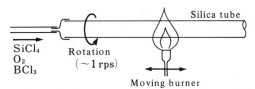

Fig. 2.8   Modified chemical vapor deposition (MCVD) method.

The tube is then "collapsed" so that no room remains inside; thus, a glass rod named "preform" is formed. The optical fiber is then drawn from this preform.

With the MCVD method, a high-purity silica core may be obtained; however, the method is tedious and time consuming. The process just described is called the "inner deposition method." In some cases, vapor deposition is performed at the surface of a silica rod; this method is called the "outer deposition method."

To raise the refractive index, usually GeO_2, TiO_2, or P_2O_5 is used as the dopant. When a complicated refractive-index profile is to be obtained, dopant contents are computer controlled on a real-time basis.

References

1. S. A. Schelkunoff, "Electromagnetic Fields." Ginn (Blaisdell), Boston, Massachusetts, 1963.
2. R. E. Collin, "Foundations for Microwave Engineering." McGraw Hill, New York, 1966.
3. W. K. H. Panofsky and M. Philips, "Classical Electricity and Magnetism." Addison-Wesley, Reading, Massachusetts, 1961.

4. M. Born and E. Wolf, "Principles of Optics," 4th ed., pp. 110–124. Pergamon, Oxford, 1970.
5. D. Marcuse, "Light Transmission Optics," Bell Laboratories Series. Van Nostrand Reinhold, New York, 1972.
6. D. Marcuse, "Theory of Dielectric Optical Waveguides." Academic Press, New York, 1974.
7. T. Izawa, Lower limit of transmission loss of an optical fiber (in Japanese), *Joint Nat. Conv. Four Inst. Elec. Eng.* Paper No. 122 (October 1976).
8. M. Horiguchi and H. Osanai, Spectral losses of low-OH-content optical fibers, *Electron. Lett.* **12**, No. 12, 310–312 (1976).
9. R. D. Maurer, Glass fibers for optical communications, *Proc. IEEE* **61**, No. 4, 452–462 (1973).

3 | Ray Theory
of Optical Fibers

At present, the standard approach to wave propagation analysis in optical fibers is the wave theory described in Chapters 4 and 5. However, ray theory, which is less general, has an advantage in that the physical picture is clear and comprehensible.

The analysis of propagation characteristics based on ray theory is presented. The characteristics of uniform-core fibers are investigated. Dispersion characteristics of nonuniform-core fibers are then discussed; it is shown that when the refractive-index profile is close to quadratic (parabolic), multimode dispersion is reduced remarkably below that of uniform-core fibers.

3.1 Introduction

As suggested in Section 2.7, the propagation of an optical beam is dealt with as a ray, provided the spatial variation of the refractive index is gradual compared to λ. Ordinary optical fibers may or may not satisfy such conditions.

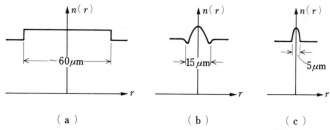

Fig. 3.1 Three typical refractive-index profiles: (a) multimode fiber, (b) graded-index (W-type) fiber, (c) single-mode fiber.

Consider three typical fiber refractive-index profiles shown in Fig. 3.1. If the wavelength is about 1 μm and the index of glass about 1.5, then the core diameters in Figs. 1a–1c will correspond to about 100, 23, and 7.5 wavelengths, respectively. In the case of Fig. 3.1b, we should also note that the fine index variation might increase the ray theory error. Thus, we may presume that ray theory

(1)   can be applied to uniform-core multimode fibers (Fig. 3.1a) with good accuracy,

(2)   can be applied to relatively thin graded-core multimode fibers, but the error might be intolerable,*

(3)   cannot be applied to single-mode fibers.

Because of restrictions (2) and (3), the standard approach to optical fiber analysis is the Maxwellian wave theory, which is applicable to any shape of the index profile. However, ray theory has an advantage in that it allows understanding of the physical picture. The ray theory analysis is presented in this chapter. The wave theory is described in Chapters 4 and 5, and the relation between the ray and wave theories is discussed in Chapter 6.

3.2   Ray Theory of Uniform-Core Fibers

3.2.1   Ray Classification in an Optical Fiber

Consider a fiber consisting of a uniform core with radius a and refractive index n_1, and cladding with refractive index n_2 where $n_2 < n_1$. The rays that can propagate in such a structure may be classified into two kinds.

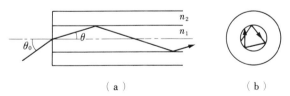

(a)   (b)

Fig. 3.2   (a) A meridional ray and its launching. (b) Concept of the skew ray.

(1)   *Meridional rays* travel in a plane that includes the fiber axis, passing it twice during one period of undulation (Fig. 3.2a).

---

* The extention of ray theory for its application to such cases [1] is investigated.

(2)   *Skew rays*, as observed from the fiber end, never pass the fiber axis; a typical path is shown in Fig. 3.2b.

3.2.2   Analysis of Meridional Rays

We consider that a ray is incident on the center of an end of a uniform-core fiber with angle θ_0 as shown in Fig. 3.2a. It enters the fiber, after refraction at the air–core boundary, with the angle θ given by Snell's law as

$$\sin \theta = \sin \theta_0 / n_1, \tag{3.1}$$

where n_1 is the core index.

On the other hand, for the ray to travel in the core subject to total reflections repeatedly at the core–cladding boundary, from Eq. (2.74),

$$\sin \theta < \sqrt{n_1^2 - n_2^2} / n_1. \tag{3.2}$$

Hence, a meridional ray can travel in the fiber only when the incident angle θ_0 satisfies

$$\sin \theta_0 < \sqrt{n_1^2 - n_2^2}. \tag{3.3}$$

In the discussion of uniform-core fibers, the relative refractive-index difference defined as

$$\Delta \equiv (n_1^2 - n_2^2)/2n_1^2 \simeq (n_1 - n_2)/n_1 \tag{3.4}$$

is often used. By using this parameter, Eq. (3.3) is rewritten as

$$\sin \theta_0 < n_1 \sqrt{2\Delta}. \tag{3.5}$$

3.2.3   Numerical Aperture

In a lens system, when the angle subtended by the pupil radius against an object is denoted by α, $\sin \alpha$ is called the numerical aperture (N.A.) of the lens system. This term is often used in discussing the launch condition of optical fibers. Equation (3.3) shows that, for an efficient launch of a meridional ray onto an optical fiber, a focusing lens system (in most cases an object lens for microscope) having a numerical aperture

$$\text{N.A.} = \sqrt{n_1^2 - n_2^2} = n_1 \sqrt{2\Delta} \tag{3.6}$$

should be used. Consequently, the quantity $(n_1^2 - n_2^2)^{1/2}$ is often called the "numerical aperture of the fiber."

3.2.4   Dispersion of Meridional Rays

In future optical communication systems, information will most probably be transmitted by the pulse-code modulation (PCM) scheme. In PCM communication, the information-transmission rate is limited by pulse spread at the receiving end. Phenomena causing pulse broadening are called dispersions. These will be discussed in detail in Section 4.4.1 after studying the wave theory of optical fibers. However, the principal facts are now briefly described.

According to wave theory terminology, total dispersion consists of three factors: multimode dispersion, waveguide dispersion, and material dispersion. In multimode fibers, *multimode dispersion predominates*; the other two factors may usually be neglected. Multimode dispersion is caused by the fact that when many propagating modes are excited by an optical pulse at the launch end of a fiber, the pulse energy distributed over all modes arrives at the receiving end within a broadened time width because the group velocity is different for different modes. According to ray theory terminology, multimode dispersion is caused by the difference in the axial average velocity of rays having different paths.

In the following, multimode dispersion is computed as the difference in the arrival times for various rays. Meridional rays in a uniform-core fiber are considered. In Fig. 3.2, the ray which travels most rapid axially is that for $\theta = 0$, whereas the slowest one is that for [see Eq. (3.2)]

$$\theta = \theta_{\max} \equiv \sin^{-1}(\sqrt{n_1^2 - n_2^2}/n_1). \tag{3.7}$$

As will be discussed in Section 4.4.2, the amount of multimode dispersion is usually represented by the difference in travel time over a unit length fiber between the fastest and slowest rays. In the present case, if the transit times for a unit length for $\theta = 0$ and $\theta = \theta_c$ are denoted by t_0 and $t_{\max}$, respectively, then

$$\tau = t_{\max} - t_0 = (\sec \theta_{\max} - 1)t_0 \simeq \tfrac{1}{2}\sin^2\theta_{\max} t_0 = t_0\Delta. \tag{3.8}$$

For example, when $n_1 = 1.5$ and $\Delta = 0.01$, we obtain $t_0 = 5 \ \mu\text{s/km}$ and hence $\tau = 50$ ns/km. The fact that τ is proportional to Δ suggests that the "weakly guiding fibers" having small Δ (see Section 1.2.3.G) are best for the low-dispersion transmission.

An approximation is made implicity in the foregoing analysis; the phase shift accompanying the total reflection at the core–cladding boundary (Goos–Haenchen shift; see Section 2.9.2) is neglected. The magnitude of the

Goos–Haenchen shift [phase shift (delay) per reflection], for cases when the electric field is normal to the incident plane, is [2]

$$\phi = -2\tan^{-1}[\sqrt{\beta^2 - n_2^2 k^2}/\sqrt{n_1^2 k^2 - \beta^2}], \tag{3.9}$$

and, for cases when the electric field is parallel to the incident plane,

$$\phi = -2\tan^{-1}[(n_1/n_2)^2(\sqrt{\beta^2 - n_2^2 k^2}/\sqrt{n_1^2 k^2 - \beta^2})], \tag{3.10}$$

where

$$k = \omega/c, \tag{3.11}$$

$$\beta = n_1 k \cos\theta, \tag{3.12}$$

the last parameter β denoting the phase constant of the propagation.

3.2.5   Analysis of Skew Rays

Because the analysis of skew rays is rather tedious, only basic ray properties are considered.

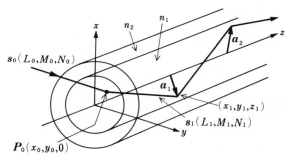

Fig. 3.3   A skew ray in a uniform-core fiber.

Consider the case shown in Fig. 3.3, in which the ray is incident on the fiber end at $\mathbf{P}_0 = x_0\mathbf{i} + y_0\mathbf{j}$ with a direction vector $\mathbf{s}_0 = L_0\mathbf{i} + M_0\mathbf{j} + N_0\mathbf{k}$, where $\mathbf{i}, \mathbf{j}$, and $\mathbf{k}$ are unit vectors [3]. If the mth reflection point is represented by a vector $\mathbf{a}_m$ and the ray direction just before the mth reflection by a direction vector $\mathbf{s}_m$, the following equation is derived from the condition that the incident and reflected rays and vector $\mathbf{a}_m$ are included in a common plane:

$$(\mathbf{s}_m - \mathbf{s}_{m+1}) \times \mathbf{a}_m = 0. \tag{3.13}$$

From the condition that the incident and reflection angles are equal to each other,

$$(\mathbf{s}_m + \mathbf{s}_{m+1}) \cdot \mathbf{a}_m = 0. \tag{3.14}$$

The condition for total reflection is

$$\mathbf{s}_m \cdot (\mathbf{a}_m/|\mathbf{a}_m|) \le (n_1^2 - n_2^2)^{1/2}/n_1. \tag{3.15}$$

After some computations described in Appendix 3A.1, the condition of Eq. (3.15) is rewritten in terms of the launch parameters,

$$[L_0^2 + M_0^2 - \{x_0 M_0 - y_0 L_0)/a\}^2]^{1/2} \le (n_1^2 - n_2^2)^{1/2}. \tag{3.16}$$

[The derivation of Eq. (3.16) given in Appendix 3A.1 seems somewhat tedious but is essentially simple analytical geometry.]

Equation (3.16) gives an important property of skew rays. When $|x_0| = a$ and $|y_0| = 0$, the left-hand side of Eq. (3.16) becomes L_0 and is independent of M_0. This means that even those rays incident almost parallel to the y axis may be confined in the core; such a ray will travel along a helical path on the core–cladding boundary, but with a very low axial velocity. Hence, if such helical rays are excited at the launch end, the uniform-core fiber may exhibit a multimode dispersion much greater than that of Eq. (3.8). Nevertheless, Eq. (3.8) remains useful as a measure of multimode dispersion because such an unfavorable launch condition is rarely encountered.

3.3 Ray Theory Analysis of Nonuniform-Core Fibers

3.3.1 Ray Classification

In nonuniform-core fibers, meridional rays and skew rays may also exist. However, in this case, skew rays do not travel along a zigzag path as shown in Fig. 3.2b but rather along more or less deformed paths (Fig. 3.4a) or even helical paths (Fig. 3.4b). Rays such as those shown in Fig. 3.4b are called helical rays.

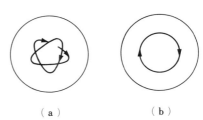

Fig. 3.4 Skew rays in a nonuniform-core fiber: (a) skew ray with complex path, (b) helical ray.

(a)   (b)

3.3.2   Basic Equations

The analysis starts from the ray equation [Eq. (2.69)] derived in Section 2.8. In cylindrical coordinates (r, θ, z), this equation is [3]

$$r \text{ component:} \quad \frac{d}{ds}\left(n\frac{dr}{ds}\right) - nr\left(\frac{d\theta}{ds}\right)^2 = \frac{dn}{dr}, \tag{3.17}$$

$$\theta \text{ component:} \quad n\left(\frac{dr}{ds}\right)\left(\frac{d\theta}{ds}\right) + \frac{d}{ds}\left(nr\frac{d\theta}{ds}\right) = 0, \tag{3.18}$$

$$z \text{ component:} \quad \frac{d}{ds}\left(n\frac{dz}{ds}\right) = 0. \tag{3.19}$$

In deriving these three expressions, an axially symmetrical and axially uniform refractive-index distribution is assumed, i.e., $dn/d\theta = 0$ and $dn/dz = 0$.

Equation (3.19) may be directly integrated to yield

$$ds = [n(r)/n_0 N_0]\,dz, \tag{3.20}$$

where N_0 is the direction cosine of the incident ray and n_0 denotes the refractive index at the incident point. Eliminating n from Eqs. (3.18) and (3.20), we obtain

$$\frac{dr}{ds}\frac{d\theta}{dz} + \frac{d}{ds}\left(r\frac{d\theta}{dz}\right) = 0, \tag{3.21}$$

which may be rewritten as

$$\frac{d}{ds}\left(r^2\frac{d\theta}{dz}\right) = 0. \tag{3.22}$$

We may integrate this equation and use the initial conditions to obtain

$$r^2\frac{d\theta}{dz} = \frac{1}{N_0}(x_0 M_0 - y_0 L_0). \tag{3.23}$$

As shown in Appendix 3A.2, by rewriting Eq. (3.17) in terms of differentiation with respect to z by using Eq. (3.20), and using Eq. (3.23), we can integrate Eq. (3.17) to obtain

$$z = \int_{r_0}^{r} N_0\left[\{n(r)/n_0\}^2 + \{(1 - (r_0/r)^2\}(x_0 M_0 - y_0 L_0) - N_0^2\right]^{-1/2}dr. \tag{3.24}$$

We can compute the path of any ray (z as a function of r) by using Eq. (3.24), if we know the refractive-index distribution $n(r)$ and the launch conditions x_0, y_0, L_0, M_0 (hence r_0 and N_0).

3.3.3   Solution Example—Meridional Rays

We first consider meridional rays. Without loss of generality, we may assume that $y_0 = M_0 = 0$ and $x_0 = r_0$. For such cases Eq. (3.24) is simplified as

$$z = \int_{r_0}^{r} N_0 [\{n(r)/n_0\}^2 - N_0^2]^{-1/2}\, dr. \tag{3.25}$$

We assume that the index distribution (profile) is quadratic,

$$n^2(r) = n_1^2 [1 - (\alpha r)^2]. \tag{3.26}$$

Substituting Eq. (3.26) into (3.25) and integrating, we obtain, as shown in Appendix 3A.3,

$$r = C \sin(\alpha n_1 z/n_0 N_0 + \psi) \qquad (\psi = \text{const}), \tag{3.27}$$

where

$$C = \alpha^{-1}[1 - N_0^2(1 - \alpha^2 r_0^2)]^{1/2}. \tag{3.28}$$

Equation (3.27) expresses an undulating ray as shown in Fig. 3.5a, whose period length is

$$\Lambda = (2\pi N_0/\alpha)[1 - (\alpha r_0)^2]^{1/2} \simeq 2\pi/\alpha. \tag{3.29}$$

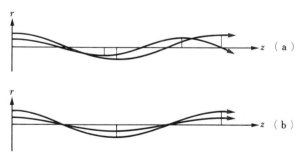

Fig. 3.5   Dependence of the length of period of undulation of meridional rays on the launching condition for (a) parabolic (quadratic) profile and (b) the profile of Eq. (3.39).

It is thus found that the period length Λ is determined solely by α in the first-order approximation, but is affected slightly by the launch conditions r_0 and N_0.

Kawakami and Nishizawa showed that Λ becomes constant for a certain index profile which is close to, but slightly deviated from, the quadratic distribution, and for that profile the average axial velocity of the ray becomes independent of the launch conditions [4, 5]. The latter property results in a zero multimode dispersion for meridional rays. We now derive this index profile from the constant axial velocity conditions.

Consider a straight ray R_0 along the axis and an undulating ray R_A, as shown in Fig. 3.6. If the maximum radial position is denoted by r_0, after computations described in Appendix 3A.4, the average axial velocity of R_A is given as

$$v_{av} = \frac{4\int_0^{r_0} (dz/dr)\, dr}{4\int_0^{r_0} (dt/dr)\, dr} = \frac{\int_0^{r_0} \beta[k^2 n^2(r) - \beta^2]^{-1/2}\, dr}{\int_0^{r_0} kn^2(r) c^{-1}[k^2 n^2(r) - \beta^2]^{-1/2}\, dr}, \qquad (3.30)$$

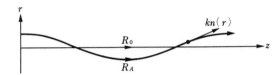

Fig. 3.6   Derivation of the optimum-index profile for meridional rays.

where c is the velocity of light and $[k^2 n^2(r) - \beta^2]^{-1/2}$ denotes the inverse of the wave number in the radial direction. On the other hand, the axial velocity of R_0 is equal to c/n_1, where $n_1 = n(0)$. Hence, we should look for an index profile for which the average axial velocity

$$v_{av} = c/n_1, \qquad (3.31)$$

regardless of the axial phase constant β, i.e., regardless of the mode.

We first note that the ray becomes parallel to the axis at $r = r_0$. Since β is conserved along the ray path as shown in the latter half of Appendix 3A.4,

$$\beta = kn(r_0). \qquad (3.32)$$

Substituting this equation into Eq. (3.30), we obtain

$$v_{av} = -c(\partial I/\partial \beta)/(\partial I/\partial k), \qquad (3.33)$$

where

$$I = \int_0^{r_0} [k^2 n^2(r) - \beta^2]^{1/2}\, dr = k \int_0^{r_0} [n^2(r) - n^2(r_0)]^{1/2}\, dr. \qquad (3.34)$$

From Eqs. (3.31) and (3.33), the differential equation

$$(\partial I/\partial k) + n_1\, \partial I/\partial \beta = 0 \qquad (3.35)$$

is derived; its solution is

$$I = f(kn_1 - \beta) \qquad (3.36)$$

On the other hand, by expressing the integrand of Eq. (3.34) as a power series in terms of $[n(r) - n(r_0)]$ and integrating, we obtain

$$I = k \sum_{m=0}^{\infty} C_m[n_1 - n(r_0)]^m = \sum_{m=0}^{\infty} C_m k^{1-m}(kn_1 - \beta)^m. \qquad (3.37)$$

For Eq. (3.37) to agree with Eq. (3.36), $C_m = 0$ for $m \neq 1$; hence, the index profile must satisfy

$$\int_0^{r_0} [n^2(r) - n^2(r_0)]^{1/2} \, dr = C_1[n_1 - n(r_0)]. \qquad (3.38)$$

If we express $n^2(r)$ by a power series of r and determine the coefficients so that Eq. (3.38) is satisfied, after computations described in Appendix 3A.5, we obtain

$$n^2(r) = n_1^2[1 - (\alpha r)^2 + \tfrac{2}{3}(\alpha r)^4 + \cdots] = n_1^2 \, \mathrm{sech}^2(\alpha r) \qquad (\alpha = \pi/2C_1). \quad (3.39)$$

This is the ideal index profile giving zero multimode dispersion for meridional rays.

That the index profile of Eq. (3.39) gives a constant Λ regardless of the launch conditions is proved as follows. Substituting Eq. (3.39) into (3.25), we can compute the integration in Eq. (3.25) to obtain

$$\sinh(\alpha r) = C' \sin(\alpha z) + \sinh(\alpha r_0), \qquad (3.40)$$

where

$$C' = [(\cosh^2(\alpha r_0)/N_0) - 1]^{1/2}. \qquad (3.41)$$

Equation (3.40) gives an undulating path for which the period length is expressed as

$$\Lambda = 2\pi/\alpha. \qquad (3.42)$$

This is independent of the launch condition as illustrated in Fig. 3.5.

3.3.4   Solution Example—Helical Rays

Next we consider helical rays which correspond to special cases of skew rays. It can be shown that, also for the helical rays, there exists an index profile that minimizes multimode dispersion [5].

Consider a helical ray traveling on a cylindrical surface of radius r, with a slant angle ϕ as shown in Fig. 3.7. If we assume beforehand in the ray equation [Eq. (2.68)] that the ray path is restricted to a cylindrical surface $r = r_0$, we may write

$$n(r) \, ds/ds = \nabla n(r), \qquad (3.43)$$

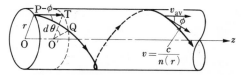

Fig. 3.7   A helical ray in a quadratic fiber (after Kawakami and Nishizawa [5]).

where **s** denotes a unit vector along the path. The modulus of the increment of **s** is, in terms of quantities shown in Fig. 3.7,

$$|ds| = \sin \phi \, d\theta. \tag{3.44}$$

On the other hand,

$$\sin \phi \, ds = TQ = r \, d\theta. \tag{3.45}$$

Hence, from Eqs. (3.43)–(3.45), we obtain

$$n(r)(\sin^2\phi)/r = -dn(r)/dr, \tag{3.46}$$

which gives the relation between the radius of the helical path r and the slant angle ϕ.

The axial velocity v_z is, from Fig. 3.7 and Eq. (3.46),

$$v_z(r) = \frac{\cos \phi}{n(r)\sqrt{\varepsilon_0\mu_0}} = \frac{[1 + (r/n)\, dn/dr]^{1/2}}{n\sqrt{\varepsilon_0\mu_0}}. \tag{3.47}$$

Equating (3.47) to a constant velocity

$$v_{z0} = 1/n_1\sqrt{\varepsilon_0\mu_0} \tag{3.48}$$

and solving the differential equation as shown in Appendix 3A.6, we obtain the solution as

$$n^2(r) = n_1^2/[1 + (\alpha r)^2] = n_1^2[1 - (\alpha r)^2 + (\alpha r)^4 - \cdots]. \tag{3.49}$$

We now conclude that when the index profile $n(r)$ satisfies Eq. (3.49), the velocities of all helical rays coincide, resulting in zero multimode dispersion for helical rays.

3.4   Summary

Note that the index profiles given by Eqs. (3.39) and (3.49) do not agree. In other words, an index profile which makes the dispersion zero for both meridional and helical rays does not exist. However, Eq. (3.39) agrees with

(3.49) within the second term of the series expansion. This suggests that the quadratic (parabolic) index profile is excellent for reducing dispersion.

However, in actual fibers, the quadratic distribution cannot be extended to an infinite radial distance but must be truncated at a certain radius, because the refractive index is usually uniform in the cladding. Therefore, the foregoing discussions are more or less idealized. The advantage of the quadratic profile, as well as its limitation, will be elucidated in Chapter 5 by more strict analyses based on wave theory.

References

1. Discussions at Workshop at Lannion, France, held jointly with the *Eur. Conf. Opt. Fiber Commun. 2nd, Paris* (September 1976).
2. D. Marcuse, "Theory of Dielectric Optical Waveguides." Academic Press, New York, 1974.
3. D. B. Keck, "Fundamentals of Optical Fiber Communications" (M. K. Barnoski, ed.), Chapter 1. Academic Press, New York, 1976.
4. S. Kawakami and J. Nishizawa, Kinetics of an optical wave packet in a lens-like medium, *J. Appl. Phys.* **38**, No. 12, 4807–4810 (1967).
5. S. Kawakami and J. Nishizawa, An optical waveguide with the optimum distribution of the refractive index with reference to waveform distorsion, *IEEE Trans. Microwave Theory Tech.* **MTT-16**, No. 10, 814–818 (1968).

4 | Wave Theory of Uniform-Core Fibers

The analysis of uniform-core fibers based on wave theory is presented.

We derive from Maxwell's equations the wave equations in cylindrical coordinates. These equations are then applied to the core and the cladding regions, and separate solutions are obtained for these two regions. However, these solutions contain some unknown parameters. From the continuity conditions of the electric and magnetic fields at the core–cladding boundary, the propagation constant of a mode as well as other unknown parameters in the solution are determined. The solutions are classified into the transverse magnetic (TM), transverse electric (TE), and hybrid (EH and HE) modes.

Important characteristics of uniform-core fibers such as the dispersion relation, cutoff conditions, LP-mode designations, field distribution, and group delay are discussed on the basis of the solution obtained.

4.1 Introduction

The concept of a "mode" is the most basic one in the wave theory of optical fibers. The reader is assumed to be familiar with this concept. As stated in Section 2.4.4, the electromagnetic wave propagating in an optical fiber consists of three kinds of modes. In addition to the TE and TM modes found in ordinary metallic waveguides, optical fibers have the so-called hybrid modes which have both axial electric and magnetic fields E_z and H_z. The hybrid modes are further classified into EH and HE modes. Thus, the exact analysis of the field and transmission characteristics of optical fibers is much more difficult than for metallic waveguides.

However, when the refractive-index difference between core and cladding of a uniform-core fiber is relatively small, we may approximate the wave equation and perform the analysis in a considerably simpler way. As stated in Section 1.2.3.G, optical fibers having small relative index differences

($\Delta \lesssim 0.01$) are called *weakly guiding fibers* [1]. (Most practical optical fibers for communication uses are weakly guiding fibers.) The field solutions for weakly guiding fibers can be approximated rather accurately.

Wave equations in cylindrical coordinates are derived first from Maxwell's equations. These wave equations are then solved for uniform-core fibers. The classification of the types of the solution leads to the concepts of TM, TE, EH, and HE modes. It will also be shown that, under the weakly guiding approximation, some modes degenerate with each other; the concept of the linearly polarized (LP) mode is derived from this fact. Practical characteristics of fibers such as dispersion and group delay are also described. Characteristics of nonuniform-core fibers will be discussed in Chapter 5.

4.2   Derivation of Basic Equations

4.2.1   Wave Equations in Cartesian Coordinates

When we consider an electromagnetic wave having angular frequency ω and propagating in the z direction with phase constant β, the electric and magnetic fields can be expressed as

$$\mathbf{E} = \text{Re}\{\mathbf{E}_0(x, y) \exp[j(\omega t - \beta z)]\}, \tag{4.1}$$

$$\mathbf{H} = \text{Re}\{\mathbf{H}_0(x, y) \exp[j(\omega t - \beta z)]\}. \tag{4.2}$$

For the Cartesian components of $\mathbf{E}_0$ and $\mathbf{H}_0$, Maxwell's equations become

$$\frac{\partial H_z}{\partial y} + j\beta H_y = j\omega\varepsilon E_x, \tag{4.3}$$

$$-j\beta H_x - \frac{\partial H_z}{\partial x} = j\omega\varepsilon E_y, \tag{4.4}$$

$$\frac{\partial H_y}{\partial x} - \frac{\partial H_x}{\partial y} = j\omega\varepsilon E_z, \tag{4.5}$$

$$\frac{\partial E_z}{\partial y} + j\beta E_y = -j\omega\mu H_x, \tag{4.6}$$

$$j\beta E_x + \frac{\partial E_z}{\partial x} = j\omega\mu H_y, \tag{4.7}$$

$$\frac{\partial E_y}{\partial x} - \frac{\partial E_x}{\partial y} = -j\omega\mu H_z. \tag{4.8}$$

From Eqs. (4.3), (4.4), (4.6), and (4.7), components E_x, E_y, H_x, and H_y are all expressed in terms of E_z and H_z as

$$E_x = -\frac{j}{\beta_t^2}\left(\beta\frac{\partial E_z}{\partial x} + \omega\mu\frac{\partial H_z}{\partial y}\right), \tag{4.9}$$

$$E_y = -\frac{j}{\beta_t^2}\left(\beta\frac{\partial E_z}{\partial y} - \omega\mu\frac{\partial H_z}{\partial x}\right), \tag{4.10}$$

$$H_x = -\frac{j}{\beta_t^2}\left(\beta\frac{\partial H_z}{\partial x} - \omega\varepsilon\frac{\partial E_z}{\partial y}\right), \tag{4.11}$$

$$H_y = -\frac{j}{\beta_t^2}\left(\beta\frac{\partial H_z}{\partial y} + \omega\varepsilon\frac{\partial E_z}{\partial x}\right), \tag{4.12}$$

where

$$\beta_t^2 = k^2 n^2 - \beta^2, \tag{4.13}$$

$$k^2 = \omega^2\varepsilon_0\mu_0, \tag{4.14}$$

and n denotes the refractive index of the medium. Parameter β_t is called the transverse phase constant or transverse wave number.

Substituting Eqs. (4.11) and (4.12) into (4.5), we obtain

$$\frac{\partial^2 E_z}{\partial x^2} + \frac{\partial^2 E_z}{\partial y^2} + \beta_t^2 E_z = 0, \tag{4.15}$$

and further by substituting Eqs. (4.9) and (4.10) into Eq. (4.8), we get

$$\frac{\partial^2 H_z}{\partial x^2} + \frac{\partial^2 H_z}{\partial y^2} + \beta_t^2 H_z = 0. \tag{4.16}$$

4.2.2 Wave Equations in Cylindrical Coordinates

For the analysis of wave propagation in optical fibers which are axially symmetric, the wave equations are rewritten in terms of cylindrical coordinates [2]. After somewhat tedious algebraic computations outlined in Appendix 4A.1, we obtain equations corresponding to Eqs. (4.9)–(4.12), (4.15), and (4.16):

$$E_r = -\frac{j}{\beta_t^2}\left(\beta\frac{\partial E_z}{\partial r} + \omega\mu\frac{1}{r}\frac{\partial H_z}{\partial\theta}\right), \tag{4.17}$$

$$E_\theta = -\frac{j}{\beta_t^2}\left(\beta\frac{1}{r}\frac{\partial E_z}{\partial\theta} - \omega\mu\frac{\partial H_z}{\partial r}\right), \tag{4.18}$$

$$H_r = -\frac{j}{\beta_t^2}\left(\beta\frac{\partial H_z}{\partial r} - \omega\varepsilon\frac{1}{r}\frac{\partial E_z}{\partial \theta}\right), \tag{4.19}$$

$$H_\theta = -\frac{j}{\beta_t^2}\left(\beta\frac{1}{r}\frac{\partial H_z}{\partial \theta} + \omega\varepsilon\frac{\partial E_z}{\partial r}\right), \tag{4.20}$$

$$\frac{\partial^2 E_z}{\partial r^2} + \frac{1}{r}\frac{\partial E_z}{\partial r} + \frac{1}{r^2}\frac{\partial^2 E_z}{\partial \theta^2} + \beta_t^2 E_z = 0, \tag{4.21}$$

$$\frac{\partial^2 H_z}{\partial r^2} + \frac{1}{r}\frac{\partial H_z}{\partial r} + \frac{1}{r^2}\frac{\partial^2 H_z}{\partial \theta^2} + \beta_t^2 H_z = 0. \tag{4.22}$$

These equations constitute the wave equations in cylindrical coordinates.

4.2.3   Separation of Variables

When we consider wave phenomena in the cylindrical coordinates, we first solve the differential equations (4.21) and (4.22) to obtain the axial components E_z and H_z, and substitute these components into Eqs. (4.17)–(4.20) to obtain other transverse components.

General solutions of Eqs. (4.21) and (4.22) are well known. We consider the axial electric field (or magnetic field) and express it as

$$E_z \text{ (or } H_z) = R_z(r)\Theta_z(\theta). \tag{4.23}$$

Substituting Eq. (4.23) into (4.21), we find that the separation of variables is possible and Eq. (4.21) yields

$$\frac{\partial^2\Theta_z}{\partial\theta^2} + n^2\Theta_z = 0, \tag{4.24}$$

$$\frac{\partial^2 R_z}{\partial r^2} + \frac{1}{r}\frac{\partial R_z}{\partial r} + \left(\beta_t^2 - \frac{n^2}{r^2}\right)R_z = 0. \tag{4.25}$$

The second equation is known as Bessel's differential equation. The solutions of these two equations are

$$\Theta_z(\theta) = \begin{cases}\cos(n\theta + \varphi) \\ \sin(n\theta + \varphi)\end{cases} \quad (\varphi = \text{const}), \tag{4.26} \atop \tag{4.27}$$

$$R_z(r) = \begin{cases}AJ_n(\beta_t r) + A'N_n(\beta_t r) & \text{(when } \beta_t \text{ is real)}, \\ CK_n(|\beta_t|r) + C'I_n(|\beta_t|r) & \text{(when } \beta_t \text{ is imaginary)}.\end{cases} \tag{4.28} \atop \tag{4.29}$$

In these solutions, n is the azimuthal mode number, A, A', C, and C' are arbitrary constants, J_n, N_n, K_n, and I_n denote the nth order Bessel functions

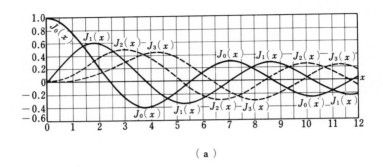

(a)

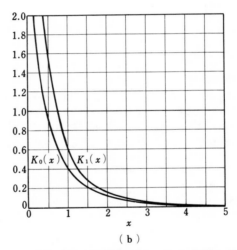

(b)

Fig. 4.1   Graphs of Bessel and modified Bessel functions: (a) functions $J_0(x)$, $J_1(x)$, $J_2(x)$, and $J_3(x)$; (b) functions $K_0(x)$ and $K_1(x)$.

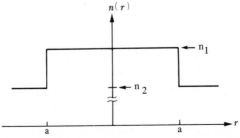

Fig. 4.2   Refractive-index profile of a uniform-core fiber.

of the first and second kinds and the nth-order modified Bessel functions of the first and second kinds, respectively. The constant rotation φ in Eqs. (4.26) and (4.27) will hereafter be omitted for simplicity.

Graphs of J_n and K_n are shown in Fig. 4.1. The Bessel functions J_n are oscillatory (and gradually damping) with respect to r, whereas the modified Bessel functions K_n decay exponentially with r. The functions N_n and I_n are not shown because these are not significant in the theory of uniform-core fibers as shown in the following section.

4.3   Wave Phenomena in Uniform-Core Fibers

4.3.1   Electromagnetic Fields in Core and Cladding

Consider a uniform-core fiber as shown in Fig. 4.2. The refractive indices in the core and cladding and core diameter are denoted by n_1, n_2, and a, respectively, as in Chapter 3.

Field solutions are first obtained independently for the core and cladding regions and are connected later at the core–cladding boundary so as to satisfy boundary conditions. However, two parameters should be given for each mode throughout the two regions: azimuthal mode number n and axial phase constant β.

We consider first the cladding solution. For light energy to propagate axially in a uniform-core fiber being confined radially, the radial field function $R_z(r)$ must be of the form of Eq. (4.29) because it must essentially be an exponentially decaying function (see Fig. 2.3). Furthermore, the second term of Eq. (4.29) should be discarded ($C' = 0$) because I_n diverges for $r \to \infty$.

In the core region, solutions proportional to N_n or K_n cannot be present because N_n and K_n both diverge on the axis ($r = 0$). The possibility of the solution proportional to I_n is also denied as follows. If $R_z(r) \propto I_n$ in the core and $R_z(r) \propto K_n$ in the cladding, these two field functions can never be connected smoothly at the core–cladding boundary as required by the boundary conditions [see Eqs. (4.51)–(4.56)], because I'_n/I_n is positive everywhere (where the prime denotes a differentiation), whereas K'_n/K_n is negative everywhere as shown in Fig. 4.1b.

Thus, $R_z(r)$ must be proportional to J_n in the core and proportional to K_n in the cladding. This means that [see Eqs. (4.28) and (4.29)] β_t must be real in the core and imaginary in the cladding, and it follows from Eq. (4.13) that

$$kn_2 < \beta < kn_1. \tag{4.30}$$

Let us temporarily assume that two sets of solutions (TM-type and TE-type solutions) are present in which

$$E_z = \begin{cases} AJ_n(\beta_{t1}r)\sin n\theta & \text{(for } r \le a), & (4.31) \\ CK_n(|\beta_{t2}|r)\sin n\theta & \text{(for } r \ge a), & (4.32) \end{cases}$$

$$H_z = 0 \qquad \text{(throughout)}, \qquad (4.33)$$

and

$$H_z = \begin{cases} BJ_n(\beta_{t1}r)\cos n\theta & \text{(for } r \le a), & (4.34) \\ DK_n(|\beta_{t2}|r)\cos n\theta & \text{(for } r \ge a), & (4.35) \end{cases}$$

$$E_z = 0 \qquad \text{(throughout)}, \qquad (4.36)$$

where suffixes 1 and 2 denote the core and cladding regions, respectively. Note that the azimuthal field function is assumed to be $\sin(n\theta)$ in the first set and $\cos(n\theta)$ in the second set; this assumption facilitates the connection of the fields at the core–cladding boundary in the cases of "hybrid modes" to be discussed in the following.

In a hollow metallic cylindrical waveguide, solutions such as the former set [Eqs. (4.31)–(4.33)] are called TM modes and those such as the latter set [Eqs. (4.34)–(4.36)] are called TE modes, as described in Section 2.4.3. In optical fibers, the TE and TM modes can still exist, but only for $n = 0$. When $n \ne 0$, boundary conditions at the core–cladding interface can be satisfied only when we choose a linear combination of TE and TM modes in the core and another linear combination (with different weighting) of TE and TM modes in the cladding, as will be shown in Sections 4.3.2 and 4.3.3. Such composite modes are called "hybrid modes."

The general expressions of the field components incorporating all the TM, TE, and hybrid modes are derived by combining Eqs. (4.31)–(4.36) and using Eqs. (4.17)–(4.20).

In the core region

$$E_z = AJ_n\left(\frac{ur}{a}\right)\sin n\theta, \qquad (4.37)$$

$$E_r = \left[-A\frac{j\beta}{u/a}J_n'\left(\frac{ur}{a}\right) + B\frac{j\omega\mu_0}{(u/a)^2}\frac{n}{r}J_n\left(\frac{ur}{a}\right) \right]\sin n\theta, \qquad (4.38)$$

$$E_\theta = \left[-A\frac{j\beta}{(u/a)^2}\frac{n}{r}J_n\left(\frac{ur}{a}\right) + B\frac{j\omega\mu_0}{u/a}J_n'\left(\frac{ur}{a}\right) \right]\cos n\theta, \qquad (4.39)$$

$$H_z = BJ_n\left(\frac{ur}{a}\right)\cos n\theta, \qquad (4.40)$$

$$H_r = \left[A \frac{j\omega\varepsilon_1}{(u/a)^2} \frac{n}{r} J_n\left(\frac{ur}{a}\right) - B \frac{j\beta}{u/a} J'_n\left(\frac{ur}{a}\right) \right] \cos n\theta, \qquad (4.41)$$

$$H_\theta = \left[-A \frac{j\omega\varepsilon_1}{u/a} J'_n\left(\frac{ur}{a}\right) + B \frac{j\beta}{(u/a)^2} \frac{n}{r} J_n\left(\frac{ur}{a}\right) \right] \sin n\theta, \qquad (4.42)$$

where

$$u = \beta_{t1}a = (k^2 n_1^2 - \beta^2)^{1/2}a, \qquad (4.43)$$

and is the normalized transverse phase constant* in the core region.

In the cladding region

$$E_z = CK_n\left(\frac{wr}{a}\right) \sin n\theta, \qquad (4.44)$$

$$E_r = \left[C \frac{j\beta}{w/a} K'_n\left(\frac{wr}{a}\right) - D \frac{j\omega\mu_0}{(w/a)^2} \frac{n}{r} K_n\left(\frac{wr}{a}\right) \right] \sin n\theta, \qquad (4.45)$$

$$E_\theta = \left[C \frac{j\beta}{(w/a)^2} \frac{n}{r} K_n\left(\frac{wr}{a}\right) - D \frac{j\omega\mu_0}{w/a} K'_n\left(\frac{wr}{a}\right) \right] \cos n\theta, \qquad (4.46)$$

$$H_z = DK_n\left(\frac{wr}{a}\right) \cos n\theta, \qquad (4.47)$$

$$H_r = \left[-C \frac{j\omega\varepsilon_2}{(w/a)^2} \frac{n}{r} K_n\left(\frac{wr}{a}\right) + D \frac{j\beta}{w/a} K'_n\left(\frac{wr}{a}\right) \right] \cos n\theta, \qquad (4.48)$$

$$H_\theta = \left[C \frac{j\omega\varepsilon_2}{w/a} K'_n\left(\frac{wr}{a}\right) - D \frac{j\beta}{(w/a)^2} \frac{n}{r} K_n\left(\frac{wr}{a}\right) \right] \sin n\theta, \qquad (4.49)$$

where

$$w = |\beta_{t2}|a = (\beta^2 - k^2 n_2^2)^{1/2}a, \qquad (4.50)$$

and is the normalized transverse attenuation constant* in the cladding region.

4.3.2   Mode Classification

Modes are first classified according to azimuthal mode number n, next by the TE, TM, and hybrid (EH and HE) mode designations, and finally by radial mode number. Classification by n and TE, TM, and hybrid mode

---

* The transverse phase and attentuation constants are often (customarily) called the normalized *propagation* constant.

designations is described. The radial mode numbering is explained in Section 4.3.6.

A. Cases n = 0

In such cases, all terms in Eqs. (4.37)–(4.42) and (4.44)–(4.49) including n vanish, and the solutions are separated into two independent groups.

(1) If we assume $B = D = 0$, modes are obtained for which only E_z, E_r, and H_θ are present, whereas $E_\theta = H_r = H_z = 0$. Such modes are called TM (transverse magnetic) modes because H_z is absent. (When the radial mode number is l, the mode is called the TM_{0l} mode.)

(2) If we assume $A = C = 0$, modes are obtained for which only H_z, H_r, and E_θ are present, whereas $H_\theta = E_r = E_z = 0$. Such modes are called TE (transverse electric) modes because E_z is absent. (When the radial mode number is l, the mode is called the TE_{0l} mode.)

B. Cases n ≥ 1

In such cases, the solution cannot be separated into TE and TM modes; only the hybrid modes can satisfy the boundary conditions at the core–cladding boundary.

Hybrid modes are further classified into EH and HE modes. However, this classification will be better understood after we investigate the proper phase (propagation) constant for each mode.

4.3.3   Mode Propagation Constants—Exact Solutions

Propagation constants are determined by the boundary conditions (conditions for the continuity of fields) at the core–cladding boundary $r = a$,

$$E_z^{core} = E_z^{clad}, \tag{4.51}$$

$$E_\theta^{core} = E_\theta^{clad}, \tag{4.52}$$

$$H_z^{core} = H_z^{clad}, \tag{4.53}$$

$$H_\theta^{core} = H_\theta^{clad}, \tag{4.54}$$

$$\varepsilon_1 E_r^{core} = \varepsilon_2 E_r^{clad}, \tag{4.55}$$

$$\mu_1 H_r^{core} = \mu_2 H_r^{clad}, \tag{4.56}$$

where suffixes 1 and 2 again denote the core and cladding regions, respectively. In nonmagnetic material, $\mu_1 = \mu_2 = \mu_0$ in Eq. (4.56).

Letting $r = a$ in Eqs. (4.37)–(4.42) and (4.44)–(4.49) and substituting into Eqs. (4.51)–(4.56), we find that coefficients A, B, C, and D must satisfy

$$AJ_n(u) - CK_n(w) = 0, \tag{4.57}$$

$$BJ_n(u) - DK_n(w) = 0, \tag{4.58}$$

$$A\frac{j\beta}{(u/a)^2}\frac{nJ_n(u)}{a} - B\frac{j\omega\mu_0}{u/a}J'_n(u) + C\frac{j\beta}{(w/a)^2}\frac{nK_n(w)}{a}$$

$$- D\frac{j\omega\mu_0}{w/a}K'_n(w) = 0, \tag{4.59}$$

$$A\frac{j\omega\varepsilon_1}{u/a}J'_n(u) - B\frac{j\beta}{(u/a)^2}\frac{nJ_n(u)}{a} + C\frac{j\omega\varepsilon_2}{w/a}K'_n(w)$$

$$- D\frac{j\beta}{(w/a)^2}\frac{nK_n(w)}{a} = 0, \tag{4.60}$$

which can be unified in a matrix form:

$$[M]\begin{bmatrix} A \\ B \\ C \\ D \end{bmatrix} = 0. \tag{4.61}$$

For the so-called "nontrivial" solution to exist for the column vector $[A, B, C, D]$,

$$\det[M] = 0, \tag{4.62}$$

which, together with Eqs. (4.43) and (4.50), leads to

$$\left[\frac{J'_n(u)}{uJ_n(u)} + \frac{K'_n(w)}{wK_n(w)}\right]\left[\frac{\varepsilon_1}{\varepsilon_2}\frac{J'_n(u)}{uJ_n(u)} + \frac{K'_n(w)}{wK_n(w)}\right]$$

$$= n^2\left(\frac{1}{u^2} + \frac{1}{w^2}\right)\left(\frac{\varepsilon_1}{\varepsilon_2}\frac{1}{u^2} + \frac{1}{w^2}\right). \tag{4.63}$$

Note that this equation contains both u and w. If we solve (4.63) simultaneously with the relation between u and w,

$$u^2 + w^2 = (k^2n_1^2 - \beta^2)a^2 + (\beta^2 - k^2n_2^2)a^2 = k^2n_1^2a^2 2\Delta \tag{4.64}$$

$$[\text{where}\quad \Delta \equiv (n_1^2 - n_2^2)/2n_1^2 \simeq (n_1 - n_2)/n_1], \tag{4.65}$$

then we can determine u and w, and hence β, for each mode. Actual solutions are now given for each mode group.

A. TM *Modes*

Since $n = 0$, the right-hand side of Eq. (4.63) vanishes. Noting further that $B = D = n = 0$, we find that Eqs. (4.58) and (4.59) are trivial (meaning merely $0 = 0$), and Eqs. (4.57) and (4.60) lead to

$$\frac{\varepsilon_1}{\varepsilon_2} \frac{J_0'(u)}{uJ_0(u)} + \frac{K_0'(w)}{wK_0(w)} = 0 \qquad (4.66)$$

which is included in Eq. (4.63).

B. TE *Modes*

In this case, again $n = 0$ and the right-hand side of Eq. (4.63) is null. Since $A = C = n = 0$, Eqs. (4.57) and (4.60) are trivial, and Eqs. (4.58) and (4.59) lead to

$$\frac{J_0'(u)}{uJ_0(u)} + \frac{K_0'(w)}{wK_0(w)} = 0 \qquad (4.67)$$

which is also included in Eq. (4.63).

C. *Hybrid Modes*

When $n \geq 1$, two sets of solutions of Eqs. (4.63) and (4.64) are obtained, which correspond to EH and HE modes. However, intuitive understanding of these modes is facilitated by investigating the weakly guiding approximation as will be shown next.

4.3.4   Mode Propagation Constants—Weakly Guiding Approximation

Under the weakly guiding approximation, $(\varepsilon_1 - \varepsilon_2)/\varepsilon_1$ is assumed to be much smaller than unity; hence it follows that

(1)   Propagation constants of TM modes [Eq. (4.66)] degenerate approximately to that of the TE modes [Eq. (4.67)], and

(2)   Propagation constants of hybrid modes ($n \geq 1$) can be expressed in a much simpler form; since $\varepsilon_1 \simeq \varepsilon_2$, Eq. (4.63) is approximated to

$$\frac{J_n'(u)}{uJ_n(u)} + \frac{K_n'(w)}{wK_n(w)} = \pm n\left(\frac{1}{u^2} + \frac{1}{w^2}\right). \qquad (4.68)$$

Equation (4.67), which is now considered to hold for both TM and TE modes, may be rewritten, by using the Bessel-function formulas $J_0'(u) =$

$-J_1(u)$ and $K_0'(w) = -K_1(w)$, as

$$-J_1(u)/uJ_0(u) = K_1(w)/wK_0(w). \qquad (4.69)$$

On the other hand, Eq. (4.68) gives two sets of solutions for the positive and negative signs. When the sign is positive, after some computations outlined in Appendix 4A.2, we obtain [3]

$$-J_{n+1}(u)/uJ_n(u) = K_{n+1}(w)/wK_n(w). \qquad (4.70)$$

Those modes whose propagation constants are given as the solution to Eqs. (4.70) and (4.64) are called EH modes. When the sign is negative, similarly

$$J_{n-1}(u)/uJ_n(u) = K_{n-1}(w)/wK_n(w) \qquad (4.71)$$

is obtained, which, together with Eq. (4.64), gives a set of modes called HE modes.

The custom that the former are called EH modes and the latter HE modes was established as early as the 1930s. There is no strong reason for the terminology; the reader may take it simply as a matter of custom. Historically, microwave engineers first referred to the lowest-order mode (mode without cutoff) in a dielectric rod waveguide as the HE_{11} mode [4]. Later, designations of HE and EH modes were derived in accordance with this custom. It is worth stating, however, that in EH modes the axial magnetic field H_z is relatively strong, whereas in HE modes the axial electric field E_z is relatively strong.

4.3.5 A Unified Expression for the Propagation Constant

With the weakly guiding approximation, a unified expression for the propagation constants can be derived from Eq. (4.69) for TM and TE modes, from Eq. (4.70) for EH modes, and from Eq. (4.71) for HE modes.

We first take the inverse of both sides of Eq. (4.71) and use Eqs. (4A.11) and (4A.12) (Appendix 4A.2) to obtain

$$\frac{u\{2[(n-1)/u]J_{n-1}(u) - J_{n-2}(u)\}}{J_{n-1}(u)}$$

$$= \frac{w\{2[(n-1)/w]K_{n-1}(w) + K_{n-2}(w)\}}{K_{n-1}(w)}. \qquad (4.72)$$

This equation leads to

$$uJ_{n-2}(u)/J_{n-1}(u) = -wK_{n-2}(w)/K_{n-1}(w). \qquad (4.73)$$

The reader will easily see a resemblance among Eqs. (4.69), (4.70), and (4.73). Actually, if we define a new parameter

$$m = \begin{cases} 1 & \text{(for TM and TE modes)}, & \text{(4.74a)} \\ n+1 & \text{(for EH modes)}, & \text{(4.74b)} \\ n-1 & \text{(for HE modes)}, & \text{(4.74c)} \end{cases}$$

all three equations can be written in the unified form

$$uJ_{m-1}(u)/J_m(u) = -wK_{m-1}(w)/K_m(w). \tag{4.75}$$

In this equation, if the index of the Bessel functions is negative (as in HE_{11} modes for which $n = 1$, $m = 0$, $m - 1 < 0$), we can change the sign of the index by using the relations $J_{-i} = -J_i$ and $K_{-i} = K_i$.

The new parameter m is useful as well as physically significant. By using this parameter, various modes are analyzed in a unified manner to bring forth the concept of LP modes, which will be described in Section 4.3.8. The physical implication of the parameter m will be considered after the LP-mode concept is introduced.

4.3.6 Traditional Mode Designation and Numbering

Summarizing the foregoing discussions, we can tabulate, within the weakly guiding approximation, the proper equations for various mode groups as shown in Table 4.1 [5, 6].

Table 4.1

Traditional Mode Designation and Numbering

Mode designations	Proper equations
TM_{0l} and TE_{0l}	$uJ_0(u)/J_1(u) = -wK_0(w)/K_1(w)$
EH_{nl} $(n \geq 1)$	$uJ_n(u)/J_{n+1}(u) = -wK_n(w)/K_{n+1}(w)$
HE_{1l}	$uJ_1(u)/J_0(u) = wK_1(w)/K_0(w)$
$HE_{ne}(n \geq 2)$	$uJ_{n-2}(u)/J_{n-1}(u) = -wK_{n-2}(w)/K_{n-1}(w)$

The first mode number (suffix n) denotes, as shown in Eqs. (4.26) and (4.27), the number of the azimuthal variation of E_z or H_z. The second mode number (suffix l) means that the propagation constant of the mode is computed from the lth smallest root of u of the simultaneous proper equations shown in Table 4.1 and Eq. (4.64). (Actually, l denotes the number of the radial variation of E_z or H_z; the exact meaning of this number becomes clearer in Section 4.3.7.)

In the wave theory of optical fibers, we define

$$v = kn_1 a(2\Delta)^{1/2}. \tag{4.76}$$

This parameter is customarily called the "normalized frequency" because it is proportional to the light frequency. It is easily shown from Eqs. (4.64) and (4.76) that

$$v^2 = u^2 + w^2. \tag{4.77}$$

When the design parameters of a uniform-core fiber (n_1, a, and Δ) and the wavelength of light are given, v can be computed. Then u is given as a function of v by using the equations given in Table 4.1 and Eq. (4.77). Finally, we can compute the phase constant β, from Eqs. (4.43) and (4.76), as

$$\beta = kn_1(1 - 2\Delta u^2/v^2)^{1/2}. \tag{4.78}$$

Typically, propagation characteristics are expressed on v versus (β/k) graphs.

4.3.7 Cutoff Frequencies

As shown in Fig. 4.1b, the modified Bessel functions of the second kind K_n decay exponentially as r increases. Hence, when the parameter w is real and finite, i.e., when

$$kn_2 < \beta, \tag{4.79}$$

the electromagnetic field in the cladding decays exponentially outward, and we can presume that the electromagnetic energy is confined in the core. However, when

$$\beta = kn_2, \tag{4.80}$$

i.e., when the phase velocity becomes equal to that of a plane wave propagating in the cladding material, Eq. (4.50) leads to $w = 0$. This means that the field in the cladding does not decay outward; therefore, the energy can no longer be confined in the core. The condition described by Eq. (4.80) is called the cutoff condition, and the light frequency at which this condition is realized is called the cutoff frequency.

In optical fibers, energy transmission by a specific mode is possible at frequencies above the cutoff frequency. Transmission modes at such frequencies are called "propagation modes." On the contrary, when the frequency is below the cutoff, $\beta < kn_2$ and hence w is imaginary. An imaginary w means a spatially oscillatory solution in the cladding region; under such conditions, radiation loss increases drastically and the wave cannot propagate axially. Proper solutions of the wave equation giving such a state are called "radiation modes." However, when $\beta < kn_2$ but the difference is very small, the energy loss due to radial radiation is partly prevented by a kind of potential barrier at the core–cladding boundary, and the wave can propagate over a certain distance. Solutions of the wave equation giving such states are called "leaky modes." A more comprehensive description of these three modes will be found in Section 5.3.1.

We now return to the problem of cutoff frequencies and compute them for each mode. Since $w = 0$ and hence $u = v$ at the cutoff, the normalized cutoff frequencies v_c for LP_{ml} mode is

$$v_c = j_{(m-1)l}, \qquad (4.81a)$$

where $j_{(m-1)l}$ denotes the lth root of $J_{m-1}(x) = 0$. For conventional mode designations, from the proper equations given in Table 4.1, we obtain

$$v_c = \begin{cases} j_{0l} & \text{for} \quad TM_{0l} \text{ and } TE_{0l} \text{ modes,} & (4.81b) \\ j_{nl} & \text{for} \quad EH_{nl} \text{ modes } (n \geq 1), & (4.81c) \\ 0 \text{ and } j_{1l} & \text{for} \quad HE_{1l} \text{ modes,} & (4.81d) \\ j_{(n-2)l} & \text{for} \quad HE_{nl} \text{ modes } (n \geq 2). & (4.81e) \end{cases}$$

At cutoff frequencies, $u = v$. Therefore, comparing Eq. (4.81) and Fig. 4.1a, one can conclude that at least at the cutoff condition, the radial mode number l represents the number of radial variations of the field in the core region because E_z (or H_z) $\propto J_n(ur/a)$. Actually, the foregoing statement (that l represents the number of the radial variations) holds not only at the cutoff but also for well-confined modes, although u varies within a certain range.

4.3.8 LP Modes

Equations (4.74) and (4.75) showed that within the weakly guiding approximation, all modes characterized by a common set of m and l satisfy a common proper equation. This means that those modes are degenerate.

Gloge [1] proposed that such degenerated modes be called "LP modes," where LP means "linearly polarized." In this new mode designation, we ignore the difference in field configurations and are concerned only with the phase constant. Modes characterized by a common set of m and l, and hence having a common phase constant (within the weakly guiding approximation), are all called LP_{ml} modes regardless of their TM, TE, EH, or HE field configurations.

The general relation between the traditional (TM, TE, EH, HE) mode designations and the LP-mode designations is tabulated in Table 4.2. More detailed correspondence between those for the 10 lowest LP modes (having 10 lowest cutoff frequencies) are tabulated in Table 4.3.

Table 4.2

Relation between Traditional- and LP-Mode Designations

Mode designation		Number of degenerating modes	Proper equations
LP	Traditional		
LP_{0l} $(m = 0)$	HE_{1l}	2	$uJ_1(u)/J_0(u) = wK_1(w)/K_0(w)$
LP_{1l} $(m = 1)$	TE_{0l} TM_{0l} HE_{2l}	4	$-uJ_0(u)/J_1(u) = wK_0(w)/K_1(w)$
LP_{ml} $(m \geq 2)$	$EH_{m-1,l}$ $HE_{m+1,l}$	4	$-uJ_{m-1}(u)/J_m(u) = wK_{m-1}(w)/K_m(w)$

Table 4.3

*Relation between Traditional- and LP-
Mode Designations for the 10 Lowest LP Modes*

LP-mode designation	Traditional-mode designations and number of modes	Number of degenerating modes
LP_{01}	$HE_{11} \times 2$	2
LP_{11}	$TE_{01}, TM_{01}, HE_{21} \times 2$	4
LP_{21}	$EH_{11} \times 2, HE_{31} \times 2$	4
LP_{02}	$HE_{12} \times 2$	2
LP_{31}	$EH_{21} \times 2, HE_{41} \times 2$	4
LP_{12}	$TE_{02}, TM_{02}, HE_{22} \times 2$	4
LP_{41}	$EH_{31} \times 2, HE_{51} \times 2$	4
LP_{22}	$EH_{12} \times 2, HE_{32} \times 2$	4
LP_{03}	$HE_{13} \times 2$	2
LP_{51}	$EH_{41} \times 2, HE_{61} \times 2$	4

The "number of degenerating modes" given in these tables requires some comment. All EH and HE modes have two linearly independent components expressed by $\sin(n\theta)$ and $\cos(n\theta)$ (e.g., when $n = 1$, the vertically and horizontally polarized waves). For TM and TE modes, no such polarization is present because these are axially symmetrical modes. The number of degenerating modes is derived by such consideration.

Detailed investigation of the electromagnetic field configurations of each mode (to be described in Section 4.3.11) reveals that "distribution of the field strength in a transverse direction (i.e., E_x or E_y) is identical for those

Table 4.4

Electric field Distribution and Strength Pattern of E_x
for the Three Lowest LP Modes

LP-mode designations	Traditional designations	Electric field distribution	Intensity distribution of E_x
LP_{01}	HE_{11} $\begin{cases} n = 1 \\ m = 0 \\ l = 1 \end{cases}$		
LP_{11}	TE_{01} $\begin{cases} n = 0 \\ m = 1 \\ l = 1 \end{cases}$		
	TM_{01} $\begin{cases} n = 0 \\ m = 1 \\ l = 1 \end{cases}$		
	HE_{21} $\begin{cases} n = 2 \\ m = 1 \\ l = 1 \end{cases}$		
LP_{21}	EH_{11} $\begin{cases} n = 1 \\ m = 2 \\ l = 1 \end{cases}$		
	HE_{31} $\begin{cases} n = 3 \\ m = 2 \\ l = 1 \end{cases}$		

modes which belong to the same LP mode." The term "linearly polarized" stems from this fact. For example, the electric field distribution and the strength pattern of E_x are shown in Table 4.4 for the three lowest LP modes: LP_{01} (HE_{11}), LP_{11} (TE_{01}, TM_{01}, HE_{21}), and LP_{21} (EH_{11}, HE_{31}).

The physical implication of the mode number m is evident from Table 4.4; i.e., m denotes "the number of variations of E_x (or E_y) observed in one rotation ($\theta = 0 \sim 2\pi$)."

The concept of the LP mode is very useful in understanding, as well as in analyzing, transmission characteristics of optical fibers. However, it should be emphasized that this concept is valid only under the weakly guiding approximation ($\Delta \ll 1$). When this approximation does not hold, degenerating modes separate from each other and the LP-mode designations lose sense.

4.3.9 Dispersion Relation

Numerical solutions of the proper equations, customarily called the dispersion relation, are given for the uniform-core fiber.

As described in Section 4.3.6, we can obtain u and v by solving the proper equations in Table 4.1 and Eq. (4.77), and obtain β by using Eq. (4.78). However, in the following representation of the dispersion relation, we use, instead of β, a customarily used variable x defined as

$$x \equiv (k^2 n_1^2 - \beta^2)/(k^2 n_1^2 - k^2 n_2^2) = u^2/v^2. \tag{4.82}$$

(x will also be used in the WKB analysis described in Chapter 5.)

The relation between β and x is readily obtained as follows.

(1)   When the electromagnetic energy is well confined in the core region, i.e., at frequencies far from cutoff,

$$\beta \simeq kn_1; \tag{4.83}$$

hence,

$$x \simeq 0. \tag{4.84}$$

(2)   At cutoff frequencies,

$$\beta = kn_2; \tag{4.85}$$

hence

$$x = 1. \tag{4.86}$$

(3)   Between $x = 0$ and $x = 1$, the relation between β and x is approximately linear under the weakly guiding approximation, because Eq. (4.82) can be approximated, when $(k^2 n_1^2 - k^2 n_2^2) \ll k^2 n_1^2$, as

$$x \simeq (kn_1 - \beta)/(kn_1 - kn_2). \tag{4.87}$$

The dispersion relation for a uniform-core fiber computed numerically by using Eqs. (4.75), (4.77), and (4.82) is shown in Fig. 4.3. Numerals in this

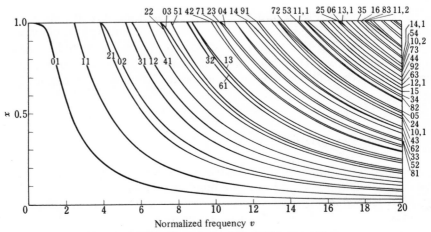

Fig. 4.3   Dispersion relation for a uniform-core fiber.

figure denote the LP-mode numbers; e.g., 01 denotes $m = 0$ and $l = 1$, 25 denotes $m = 2$ and $l = 5$. It is seen that the LP_{01} mode (HE_{11} mode) is the lowest mode and has no cutoff. The second lowest LP mode is the LP_{11} mode (consisting of TE_{01}, TM_{01}, and HE_{21} modes) whose normalized cutoff frequency is given, from Eq. (4.81a), as

$$v_c(LP_{11}) = j_{01} = 2.405. \qquad (4.88)$$

4.3.10   Single-Mode and Multimode Fibers

Optical fibers in which only one mode (actually only the HE_{11} mode) can propagate are called single-mode (or monomode) fibers. Fibers in which plural modes can propagate are called multimode fibers. The condition for the single-mode operation is obtained, from Eq. (4.88), as

$$v = kn_1 a \sqrt{2\Delta} < 2.405, \qquad (4.89)$$

which gives the maximum core diameter $2a_{max}$ or the maximum relative index difference Δ_{max} for a prescribed frequency.

For example, when $\lambda = 900$ nm (hence $k = 2\pi/\lambda \simeq 7 \times 10^6$ m^{-1}), $n = 1.5$ and $\Delta = 0.002$ the single-mode condition is given as $2a_{max} = 7.3$ μm. We should note that in hollow metallic waveguides the cutoff condition for the lowest modes is given approximately as $2a \sim \lambda$, where the symbol $\sim$ denotes "the same order of magnitude." The value $2a_{max} = 7.3$ μm is noteworthy because it is much greater than the wavelength. When the relative index difference Δ decreases, the maximum radius a_{max} increases further. This means that when the fiber is weakly guiding ($\Delta \ll 1$), the single-mode fiber can be connected without excessive technological difficulties.

In Section 3.2.4, the advantage of the weakly guiding multimode fiber in achieving a low multimode dispersion has been emphasized. In single-mode fibers, the weakly guiding design is also advantageous in the above sense. In both cases, of course, the index difference Δ cannot be reduced too much because radiation loss due to fiber bending becomes intolerable when Δ is excessively small.

4.3.11   Field Distributions

We now investigate the electromagnetic field distributions for each mode [5]. We note first that the direction of the electric field in the $r\theta$ plane is, as seen in Fig. 4.4,

$$r\,d\theta/dr = E_\theta/E_r. \tag{4.90}$$

We can compute the right-hand side of this equation for each mode, by using field-component formulas to be given in Section 5.2.4, as

$$\frac{r\,d\theta}{dr} = \begin{cases} \infty, & \text{TE}_{0l} \text{ mode,} \\ 0, & \text{TM}_{0l} \text{ mode,} \\ \tan(n\theta + \varphi_n), & \text{EH}_{nl} \text{ mode,} \\ -\tan(n\theta + \varphi_n), & \text{HE}_{nl} \text{ mode.} \end{cases} \tag{4.91}$$

These equations can be integrated to give the electric lines of force are given as

$$\begin{array}{ll} r = \text{const}, & \text{TE}_{0l} \text{ mode,} \\ \theta = \text{const}, & \text{TM}_{0l} \text{ mode,} \\ r = C|\sin(n\theta + \varphi_n)|^{1/n}, & \text{EH}_{nl} \text{ mode,} \\ r = C|\sin(n\theta + \varphi_n)|^{-1/n}, & \text{HE}_{nl} \text{ mode,} \end{array} \tag{4.92}$$

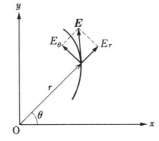

Fig. 4.4   Direction of the electric field.

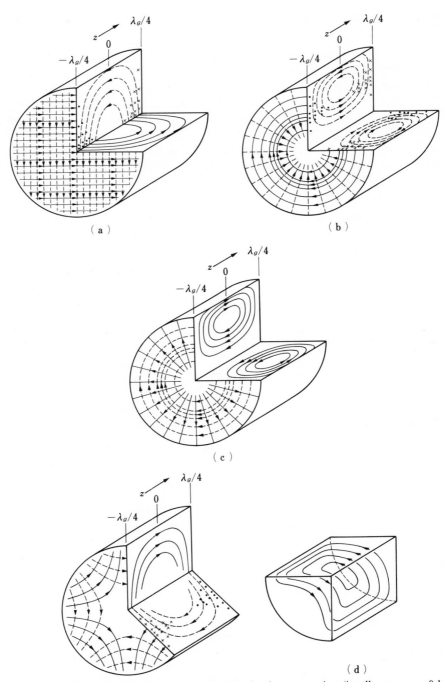

Fig. 4.5 Electromagnetic field configurations in the core region (in all cases: $x = 0.1$; E:——, H:----): (a) HE_{11} mode, (b) TE_{01} mode, (c) TM_{01} mode, (d) HE_{21} mode, (e) EH_{11} mode, and (f) HE_{31} mode.

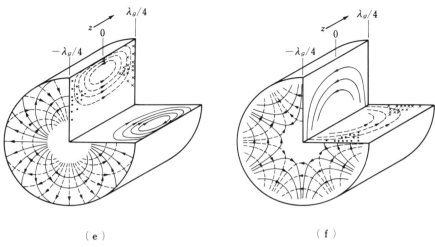

(e)　　　　　　　　　　　　　　　　　　　　(f)

Fig. 4.5 (continued)

where C is a constant. Similarly, the magnetic lines of force are expressed as

$$
\begin{aligned}
\theta &= \text{const}, & \text{TE}_{0l} \text{ mode}, \\
r &= \text{const}, & \text{TM}_{0l} \text{ mode}, \\
r &= C|\cos(n\theta + \varphi_n)|^{1/n}, & \text{EH}_{nl} \text{ mode}, \\
r &= C|\cos(n\theta + \varphi_n)|^{-1/n}, & \text{HE}_{nl} \text{ mode}.
\end{aligned}
\tag{4.93}
$$

On the other hand, the electric lines of force (ELF) and magnetic lines of force (MLF) in the rz plane are

$$
\begin{aligned}
rR(r) &= C/|\sin(\omega t - \beta z)|, & \text{MLF of TE}_{0l} \text{ modes}, \\
rR(r) &= C/|\sin(\omega t - \beta z)|, & \text{ELF of TM}_{0l} \text{ modes}, \\
r^{n+1}R(r) &= C/|\sin(\omega t - \beta z)|, & \text{ELF and MLF of EH}_{nl} \text{ modes}, \\
r^{-n+1}R(r) &= C/|\sin(\omega t - \beta z)|, & \text{ELF and MLF of HE}_{nl} \text{ modes}.
\end{aligned}
\tag{4.94}
$$

Tracing the electric and magnetic fields expressed by Eqs. (4.92)–(4.94), we find that the field configurations of TE, TM, EH, and HE modes look entirely different even when they belong to the same LP mode. However, if we investigate the strength of E_x (or E_y), or the transmission power density which is given as the real part of the axial component of the complex Poynting's vector [see Eq. (2.39)], we find that different modes belonging to the same LP mode show an (approximately) identical distribution pattern.

The field configurations of the LP_{01} mode (HE_{11} mode), LP_{11} modes (TE_{01}, TM_{01}, and HE_{21} modes), and LP_{21} modes (EH_{11} and HE_{31} modes) are shown in Fig. 4.5, in which $x = 0.1$ is assumed. The normalized power

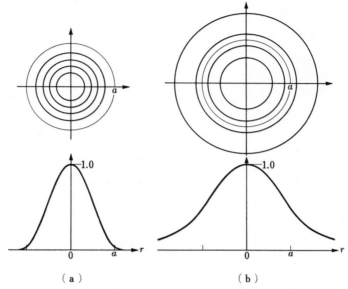

Fig. 4.6   Normalized power density distribution of the LP_{01} (HE_{11}) mode when (a) $x = 0.1$ and (b) $x = 0.9$.

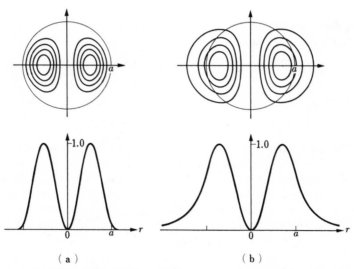

Fig. 4.7   Normalized power density distribution of the LP_{11} (TE_{01}, TM_{01}, and HE_{21}) mode when (a) $x = 0.1$ and (b) $x = 0.9$.

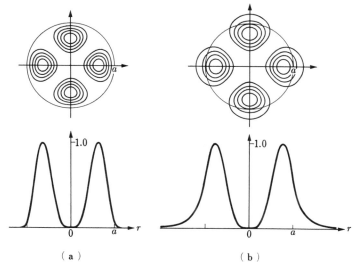

Fig. 4.8 Normalized power density distribution of the LP_{21} (EH_{11} and HE_{31}) mode when (a) $x = 0.1$ and (b) $x = 0.9$.

density for the same mode groups are shown in Figs. 4.6–4.8 for $x = 0.1$ (relatively well-confined case) and $x = 0.9$ (close-to-cutoff case). In these figures a denotes the core radius.

4.4   Dispersion Characteristics

4.4.1   Mechanisms Generating Dispersion

In optical fiber communication systems in the future, the pulse-code modulation (PCM) will be most commonly used in modulating the carrier light. In pulse-modulation schemes such as PCM, the transmitted pulse waveform is always subject to waveform distortion during its travel through the transmission medium, in the present case, the optical fiber. Waveform distortion usually results in broadening the received pulse waveform, thus limiting the transmission capacity of the optical fiber as a communications channel.

Generally speaking, the pulse-waveform distortion in a PCM system stems from the "dispersion" (difference) in the delay time (or the group velocity) of various "components" contained in the pulse. In the present case the term "component" implies

(1)   portions of light energy transmitted by various modes, and
(2)   various frequency components.

In more detail, the dispersion in an optical fiber is sorted into the following three categories according to cause.

(1)   *Multimode dispersion*   As will be shown, components of light transmitted by different modes exhibit different group velocities. In a multimode fiber, this effect causes waveform distortion, which is called multimode dispersion,* or intermodal delay difference, or delay difference between modes. This effect, of course, is absent in single-mode fibers.

(2)   *Material dispersion*   All existing fiber materials exhibit "color" dispersion, i.e., the refractive-index variation is a function of frequency (or wavelength) of light. On the other hand, the transmitted optical pulse inevitably has a finite frequency bandwidth. The delay distortion in the output-pulse waveform caused by this effect is called material dispersion.

(3)   *Waveguide dispersion*   Even though we could assume that material dispersion is entirely absent, the delay distortion due to the finite frequency bandwidth would remain because the mode group velocity in a waveguiding structure is generally a function of frequency. The delay distortion caused by this effect is called waveguide dispersion.

Category (1) should be considered only for multimode fibers, whereas dispersions due to categories (2) and (3) are present both in multimode and single-mode fibers. Dispersions due to categories (2) and (3) are both proportional to the frequency bandwidth of the transmitted pulse. Thus, categories (2) and (3) are of a similar nature and are generically called wavelength dispersions, whereas category (1) is of a somewhat different nature.

A comment should be added concerning the frequency bandwidth of the transmitted pulse. The bandwidth spread has two causes: the original spectral spread of the light source itself, and the spread of the sideband due to pulse modulation. As will be shown in Section 4.4.7, the original spectral spread is usually much broader than the spread due to modulation. For example, when a multimode semiconductor laser is used as the light source, the wavelength spectral spread is typically 2 nm, whereas the spread due to modulation, even at a modulation frequency as high as 1 GHz, is about 20 pm (see Fig. 4.12b).

---

* The term "dispersion" was originally used only for the variation of characteristics due to frequency variation. Therefore, some people dislike the term "multimode dispersion" and prefer a little lengthy but more accurate term "intermodal delay difference" or "delay difference between modes." It is principally to save space that "multimode dispersion" is used in this book.

4.4.2 Derivation of Delay-Time Formula

We define "delay time" as the "time required to propagate over a unit length (usually 1 km) of an optical fiber." This is also called "group delay." The expression for the group velocity v_g has been given in Eq. (2.33). Using this equation, the delay time (group delay) t at angular frequencies ω around ω_0 ($= 2\pi f_0$) is, within the first-order approximation,

$$t = 1/v_g = d\beta/d\omega = [d\beta/d\omega]_{\omega = \omega_0} + (\omega - \omega_0)[d^2\beta/d\omega^2]_{\omega = \omega_0}. \quad (4.95)$$

The first term in this equation gives a constant delay, but is different for each mode, thus contributing to the multimode dispersion. The second term contributes to the wavelength dispersions. Substituting the spectral spread of the transmitted pulse $2\pi\,\delta f$ into ($\omega - \omega_0$) of the second term, we obtain the sum of the material and waveguide dispersions.

For uniform-core fibers, the delay time t can be expressed in a relatively simple form by using the parameter x defined in Eq. (4.82). We obtain first from that equation

$$\beta = kn_1(1 - 2x\Delta)^{1/2}. \quad (4.96)$$

Substituting this equation into (4.95), we obtain, after some computations,

$$t = \frac{N_1}{c} \frac{[1 - \Delta(1 + y/4)Qx]}{(1 - 2x\Delta)^{1/2}}, \quad (4.97)$$

where the newly introduced parameters N_1, y, and Q are defined as follows.

Parameter N_1, which was first used by Gloge [7] and is often called the "group index," is defined (in this case for the core region) as

$$N_1 \equiv c/v_{g1} = d(kn_1)/dk = n_1 + k\,dn_1/dk, \quad (4.98)$$

where v_{g1} denotes the group velocity of a plane wave in the core material. The first term represents the constant part of the index, whereas the second term represents the dispersion of the material. As will be seen in Fig. 4.11a, the second term is usually much smaller than the first term, and an approximation

$$N_1 \simeq n_1 \quad (4.99)$$

is often permissible.

Parameter y has been proposed by Olshansky [8, 9] to express the "difference between the material dispersions in core and cladding," and is defined as

$$y = \frac{2n_1}{N_1} \frac{k}{\Delta} \frac{d\Delta}{dk} = -\frac{2n_1}{N_1} \frac{\lambda}{\Delta} \frac{d\Delta}{d\lambda}. \quad (4.100)$$

Finally, parameter Q is defined as

$$Q = (2v/u)\,du/dv. \tag{4.101}$$

For a uniform-core fiber, the relation between parameters u and v is given by Eqs. (4.75) and (4.77). Computation based on these equations leads to

$$Q = 2(1 - \xi_m)/x(1 - \xi_m/\zeta_m), \tag{4.102}$$

where

$$\xi_m(w) = K_m^2(w)/K_{m-1}(w)K_{m+1}(w), \tag{4.103}$$

$$\zeta_m(u) = J_m^2(u)/J_{m-1}(u)J_{m+1}(u). \tag{4.104}$$

We return to Eq. (4.95) and note that the delay time t includes the constant delay (the first term) as well as the dispersion (the second term). We are first concerned with the second term. Denote the pulse spread due to the second term by τ, and define a new quantity σ called "specific dispersion"; σ is expressed as

$$\tau = (\delta\lambda/\lambda)\sigma. \tag{4.105}$$

This equation tells that the pulse width τ can be computed as the product of τ and the relative bandwidth $(d\lambda/\lambda)$.

Equations (4.95) and (4.105) lead to

$$\sigma = \omega_0\,d^2\beta/d\omega^2. \tag{4.106}$$

For uniform-core fibers, using Eqs. (4.82), (4.96), and (4.98), we obtain

$$\sigma \simeq \frac{\lambda^2}{c}\frac{d^2 n_1}{d\lambda^2} + \frac{\Delta N_1 v}{c}\frac{d^2}{dv^2}\left[v(1 - x)\right]. \tag{4.107}$$

The first term in this equation expresses the material dispersion. The second term expresses the waveguide dispersion, because it is proportional to the derivative of a function of v and x with respect to v. Therefore, henceforth the first and second terms in Eq. (4.107) will be denoted by σ_m (material) and σ_w (waveguide), respectively, so that

$$\sigma = \sigma_m + \sigma_w, \tag{4.108}$$

$$\sigma_m = (\lambda^2/c)\,d^2 n_1/d\lambda^2. \tag{4.109}$$

$$\sigma_w = (\Delta N_1 v/c)\,d^2[v(1 - x)]/dv^2. \tag{4.110}$$

The delay time t and the three dispersions will be computed. The magnitudes of the three dispersions will then be compared in Section 4.4.7.

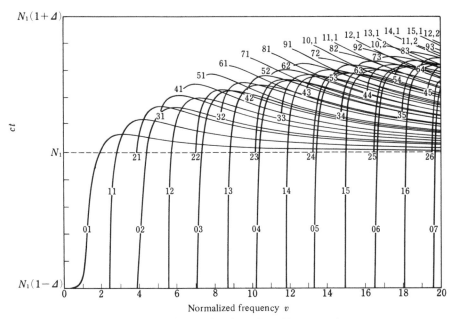

Fig. 4.9 Normalized delay time of a uniform-core fiber.

4.4.3 Computation of Delay Time

The delay time of a uniform-core fiber computed by using Eq. (4.97) is shown in Fig. 4.9 for various LP modes as functions of the normalized frequency v. In this computation, it is assumed that $N_1 = \text{const} (= n_1)$ and $y = 0$. The numerals in the figure denote (as in Fig. 4.3) the LP-mode numbers; e.g., 41 denotes $m = 4$ and $l = 1$, i.e., the LP_{41} mode. The ordinate is a dimensionless quantity $ct [= c/v_g$; see Eq. (4.95)], where c denotes the light velocity.

When the light energy is well confined in the core, x approaches zero. Therefore, in such cases, from Eq. (4.97) and the assumption that $N_1 = n_1$, we have the relation

$$ct \simeq N_1 = n_1. \tag{4.111}$$

This relation could also be understood directly from the physical meaning of N_1. Actually, at the right-hand side of Fig. 4.9, curves suggest that when v becomes large enough, ct will approach N_1. Another fact found from Fig. 4.9 is that for LP_{0l} and LP_{1l} modes, ct agrees with $N_1(1 - \Delta) \simeq N_2$ at the cutoff frequency.

In computing the dispersion relation shown in Fig. 4.9, it has been assumed that $N_1 = n_1$ and $y = 0$. Therefore, the material dispersion is evidently not considered. The material dispersion is a quantity proportional to (dN_1/dv), whereas the waveguide dispersion is proportional to the slope of the curves in Fig. 4.9, i.e., $d(ct)/dv$. Therefore, to show both dispersions on such a graph, we should distort Fig. 4.9 so that N_1 is no longer constant as shown by the horizontal dashed line, but follows the actual variation as a function of the normalized frequency v. If we do so, however, the graph is no longer universal but proper to a specific glass material.

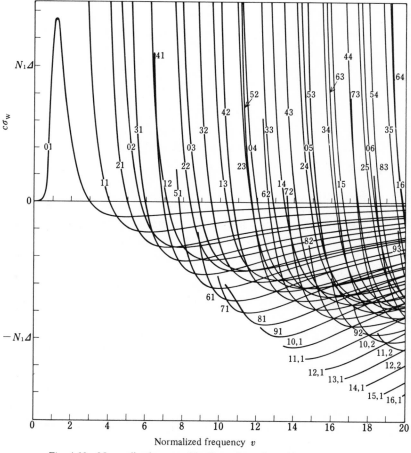

Fig. 4.10   Normalized waveguide dispersion of a uniform-core fiber.

4.4.4 Waveguide Dispersion

The waveguide dispersion σ_w of a uniform-core fiber computed by using Eq. (4.110) under the same assumption as in Fig. 4.9, $N_1 = n_1$ and $y = 0$, is shown in Fig. 4.10 for various LP modes. In this figure, the ordinate is a dimensionless quantity $c\sigma_w$. This quantity will hereafter be called the "normalized waveguide dispersion." To give the measure to the magnitude of the normalized waveguide dispersion, $+N_1\Delta$ and $-N_1\Delta$ are shown on the ordinate.

Note that σ_w is negative for all modes at well-confined conditions (at frequencies far from cutoff), but is positive at frequencies close to cutoff. Especially noteworthy is the positive waveguide dispersion of the LP_{01} (HE_{11}) mode in the single-mode frequency range. This fact is particularly important when we consider the possibility of the cancellation between the waveguide and material dispersions (see Section 7.2).

4.4.5 Material Dispersion

The material dispersion is determined by the optical characteristics of the fiber material—in most cases, glass. As an example, characteristics of typical silica core and cladding materials are shown in Fig. 4.11 [9]. The core material (refractive index n_1) is silica glass doped with TiO_2 (3.4 wt %), whereas the cladding material (refractive index n_2) is pure fused silica.

Figures 4.11a–4.11d show the characteristics as functions of wavelength: (a) refractive indices n_1, n_2 and group indices N_1 and N_2; (b) nomalized material dispersion $c\sigma_m$ $[= \lambda^2(d^2n/d\lambda^2)]$ for the core and cladding, where c denotes light velocity; (c) the relative refractive-index difference Δ; and (d) Olshansky's parameter y.

4.4.6 Multimode Dispersion

The magnitude of the multimode dispersion can be obtained from Fig. 4.9. We note that the dispersion is given as the spread of the delay time for various modes at a given frequency, i.e., the distribution of curves crossing a vertical line in Fig. 4.9.

Figure 4.9 shows that the multimode dispersion magnitude (spread of the delay time) is of the same order of magnitude as the delay-time difference between two waves propagating with group velocities c/N_1 and c/N_2, i.e., plane waves propagating in the core and cladding materials. This fact

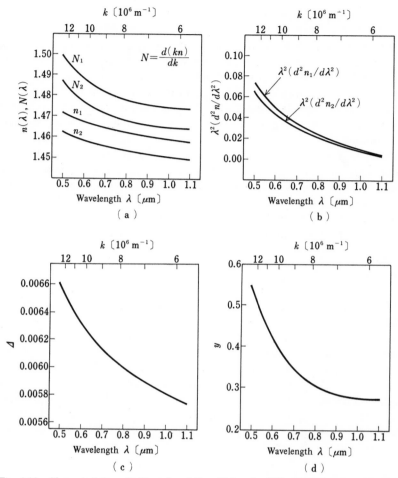

Fig. 4.11 Characteristics of silica glass (after Olshansky [9]). (a) Refractive indices and group indices of core and cladding materials. (b) Normalized material dispersion of core and cladding materials. (c) Relative refractive-index difference. (d) Olshansky's parameter.

substantiates the validity of the approximate formula [Eq. (3.8)] which has been derived by ray theory.

4.4.7   Comparison of Three Dispersions

First compare the material and waveguide dispersions in a single-mode uniform-core fiber. Figure 4.11b shows that at $\lambda = 850$ nm the normalized material dispersion $c\sigma_{\mathrm{m}} \sim 0.02$. On the other hand, Fig. 4.10 shows that when

$N_1 = 1.5$ and $\Delta = 0.002$, the maximum normalized waveguide dispersion $c\sigma_w \simeq 0.003$. Hence, in this case the waveguide dispersion is smaller than the material dispersion. This statement remains valid in most cases, except at longer wavelengths (e.g., when $\lambda > 1.0~\mu$m; see Fig. 4.11b), where the material dispersion decreases drastically.

Next consider a multimode fiber having a uniform core. In this case, as seen at the right-hand side of Fig. 4.10, the magnitude of the waveguide dispersion is of the same order of magnitude as that in the single-mode case ($c\sigma_w \sim N_1\Delta$). More precisely, the waveguide dispersion is different for each mode; its contribution to the entire delay distortion is quite complex. However, in most cases the relation $\sigma_m \gg \sigma_w$ still holds at $\lambda = 850$ nm.

In multimode fibers, material and multimode dispersions are the important factors. To compare their magnitudes, we must assume the spectral spread of the light source to be used in the system. As pointed out in Section 4.4.1, the spectral spread of the transmitted signal consists of two components: the original spectral spread of the light source itself and the spread of the sideband due to pulse modulation.

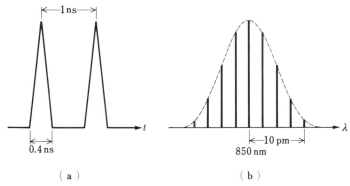

(a)   (b)

Fig. 4.12 (a) A typical high bit-rate PCM waveform. (b) Spectral spread corresponding to the waveform shown in (a).

We first compare these two components. Figure 4.12a shows a part of a typical high bit-rate PCM waveform with 1-Gbit/s repetition frequency. The corresponding spectrum is shown in Fig. 4.12b, which shows that the spectral spread is about 20 pm in wavelength. On the other hand, typical characteristics of three kinds of light sources (light emitting diode, multimode semiconductor laser, and single-mode semiconductor laser) are tabulated in the upper half of Table 4.5. It is found that for the light-emitting diode and multimode laser, the spectral spread due to pulse modulation is much smaller than the original spectral spread of the light source and could

Table 4.5

Typical Characteristics of Three Light Sources

Parameters	Light emitting diode	Multimode semiconductor laser	Single-mode semiconductor laser
λ [nm]	900	850	850
$\delta\lambda$ [nm]	30	1	10^{-4}
$\delta\lambda/\lambda$	3.3×10^{-2}	1.2×10^{-3}	1.2×10^{-7}
$c\sigma_m$	0.016	0.020	0.020
σ_m [s/m]	5.4×10^{-11}	6.7×10^{-11}	6.7×10^{-11}
τ_m [ps/km]	800	80	8×10^{-3}
$c\sigma_w \simeq$	0.003	0.003	0.003
σ_w [s/m] $\simeq$	10^{-11}	10^{-11}	10^{-11}
τ_w [ps/km] $\simeq$	330	12	1.2×10^{-3}
τ_{multi} [ns/km]	10	10	10

practically be neglected. In the case of the single-mode laser, this relation is reversed.

In the lower half of Table 4.5, material dispersion per unit distance τ_m $[= (\delta\lambda/\lambda)\sigma_m]$, waveguide dispersion per unit distance τ_w $[= (\delta\lambda/\lambda)\sigma_w$, where $c\sigma_w$ is assumed to be 0.003], and approximate magnitude of the multimode dispersion computed from Fig. 4.9 ($\Delta = 0.0012$ being assumed) are tabulated for the three typical light sources. It will be seen that in multimode fibers the multimode dispersion is predominant in all three cases.

4.5 Summary

We have studied the wave theory of uniform-core fibers. Concepts of modes, dispersion relations, cutoff conditions, field distribution, and delay characteristics have been derived. Characteristics of nonuniform-core fibers will be discussed in Chapter 5.

References

1. D. Gloge, Weakly guiding fibers, *Appl. Opt.* **10**, No. 10, 2252–2258 (1971).
2. D. Marcuse, "Light Transmission Optics," Bell Laboratories Series. Van Nostrand Reinhold, New York, 1972.

3. G. Biernson and D. J. Kinsley, Generalized plots of mode patterns in a cylindrical dielectric waveguide applied to retinal cones, *IEEE Trans.* Microwave Theory Tech. **MTT-13**, No. 5, 345–356 (1965).

4. J. R. Carson, S. P. Mead, and S. A. Schelkunoff, Hyper frequency waveguides—mathematical theory, *Bell Syst. Tech. J.* **15**, 310–333, (1936)

5. E. Snitzer, Cylindrical dielectric waveguide modes, *J. Opt. Soc. Am.* **51**, No. 5, 491–498 (1961).

6. A. W. Snyder, Asymptotic expressions for eigenfunctions and eigenvalues of a dielectric or optical waveguide, *IEEE Trans. Microwave Theory Tech.* **MTT-17**, No. 2, 1130–1138 (1969).

7. D. Gloge, Dispersion in weakly guiding fibers, *Appl. Opt.* **10**, No. 11, 2442–2445 (1971).

8. R. Olshansky *et al.*, Material effects on minimizing pulse broadening, a paper read at the *Top. Meeting Opt. Fiber Transmission*, at *Williamsburg. Virginia* (January, 1975).

9. R. Olshansky and D. B. Keck, "Pulse broadening in graded-index optical fibers, *Appl. Opt.* **15**, No. 2, 483–491 (1976).

5 | Wave Theory of Nonuniform-Core Fibers

The wave theory of nonuniform-core fibers (fibers having arbitrary refractive-index profiles) is presented by extending the wave theory of uniform-core fibers described in the preceding chapter.

The major topics discussed are (1) the relation between vectorial wave (exact) analysis and scalar wave (approximate) analysis, (2) how and under what approximations the concept of modes (both traditional and LP modes) holds for nonuniform-core fibers, and (3) the analysis of propagation characteristics of nonuniform-core fibers: WKB method, Rayleigh–Ritz method, power-series expansion method, finite element method, and staircase-approximation method.

The so-called leaky waves are discussed in connection with WKB analysis. The propagation characteristics of the so-called α-power fibers are derived by using the power-series expansion method and are investigated in some detail to give the starting point for considering the optimum refractive-index profile, discussed in Chapter 7.

5.1 Introduction

As shown in Table 4.5, the magnitude of the multimode dispersion in a uniform-core fiber is typically 10 ns/km. This means that in a 10-km-long PCM communication system, the information–transmission rate is at most 10 Mbit/s, which is not satisfactory for modern communication networks.

On the other hand, the ray theory described in Chapter 3 suggests that, at least within the ray-theory approximation, the pulse spread at the receiving end could be reduced dramatically compared with that in uniform-core fibers, if we make the refractive-index profile close to quadratic. In ray theory we consider that the pulse spread results from dispersion of the axial velocity among various rays. However, as stated in Section 3.2.4, this is nothing but multimode dispersion in the terminology of the wave theory of optical fibers.

The relation between ray theory and wave theory will be discussed in detail in Chapter 6. We now recall the following well-known facts:

(1) In a microwave rectangular waveguide in which TE_{n0} modes propagate, each TE_{n0} mode can be expressed by two skewly propagating plane waves (rays).

(2) As the mode number n increases, the angle between the ray and the z axis increases.

In optical fibers, the relation between rays and modes is essentially identical.

The fact that pulse spread can be reduced dramatically by making the refractive-index profile quadratic leads us to the expectation that if we shape the profile more freely, we might be able to reduce further intermodal pulse spread. The investigation of fibers having arbitrarily shaped index profiles has been prompted by this expectation.

Optical fibers having arbitrary index profiles are often called graded-index fibers. In this book, however, the term "nonuniform-core fibers" is used throughout in contrast to "uniform-core fibers," because we are also interested in, for example, three-layer structure which does not have a "graded" core but still has a "nonuniform" core.

Nonuniform-core fibers are further classified into those for multimode transmission and those for single-mode transmission. Several examples of the profile are shown in Fig. 5.1. In Fig. 5.1c, a valley is made at the core–cladding boundary, sometimes for reducing the internal mechanical stress due to the gradient of dopant concentration, and sometimes for reducing the multimode dispersion (see Section 5.6). The dent at the center (Figs. 5.1d and 5.1e) is often found in fibers made by the MCVD method: it is produced by the out-diffusion of the dopant during the "collapse" process (see Section 2.11.3).

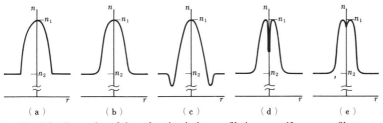

Fig. 5.1 Examples of the refractive-index profile in nonuniform-core fibers.

Various methods of the wave-theory analysis which can easily be applied to any of the profiles shown in Fig. 5.1 are presented. The methods described are

(1)   WKB method,
(2)   Rayleigh–Ritz method,
(3)   power-series expansion method,
(4)   finite element method, and
(5)   staircase-approximation method.

The so-called leaky waves are discussed in connection with WKB analysis. Actual multimode dispersion of various profiles is investigated on the basis of WKB and power-series expansion analyses. The error caused by the scalar wave approximation is discussed mainly in connection with the finite-element-method analysis and partly with the staircase-approximation analysis.

In addition to the five analytical methods mentioned, other methods such as the perturbation method and numerical integration method are usable. Descriptions of these methods, however, are omitted because of the space restriction. Comparison of the advantages and disadvantages of various methods will be found in Chapter 6.

5.2   Basic Equations and Mode Concepts in Nonuniform-Core Fibers

We first derive the basic equations for nonuniform-core fibers. Next it will be shown that in the analysis of nonuniform-core fibers, the "small index gradient" and "weakly guiding" approximations greatly simplifies the analysis by making the wave equation scalar.

In Section 5.2.4, it will be shown that the traditional mode designations (TE, TM, EH, and HE modes) are usable in the strict sense even for nonuniform-core fibers. The LP-mode concept holds only under proper approximations as for uniform-core fibers. The required approximations will also be clarified.

5.2.1   Derivation of Basic Equations

We first set cylindrical coordinates (r, θ, z) along the axis of a fiber, and express the electric and magnetic fields in the fiber by using the axial and transverse components as

$$\mathbf{E} = [\mathbf{E}_t(r, \theta) + \mathbf{k}E_z(r, \theta)]e^{j(\omega t - \beta z)}, \tag{5.1}$$

$$\mathbf{H} = [\mathbf{H}_t(r, \theta) + \mathbf{k}H_z(r, \theta)]e^{j(\omega t - \beta t)} \tag{5.2}$$

In these equations $\mathbf{E}_t$ and $\mathbf{H}_t$ denote electric and magnetic field vectors included in the $r\theta$ plane, $\mathbf{k}$ is the unit vector in the axial direction, E_z and H_z the axial field components (scalar quantities), ω the angular frequency of light, and β the phase constant in the axial direction.

Substituting Eqs. (5.1) and (5.2) into the Maxwell's equations

$$\nabla \times \mathbf{E} = -\mu \, \partial \mathbf{H}/\partial t, \tag{5.3}$$

$$\nabla \times \mathbf{H} = \varepsilon \, \partial \mathbf{E}/\partial t, \tag{5.4}$$

$$\nabla \cdot \varepsilon \mathbf{E} = 0, \tag{5.5}$$

$$\nabla \cdot \mu \mathbf{H} = 0, \tag{5.6}$$

and performing some algebraic computations described in Appendix 5A.1, we can obtain the following four differential equations in terms of $\mathbf{E}_t$, $\mathbf{H}_t$, E_z, and H_z:

$$-\mu \nabla \times (\mu^{-1} \nabla \times \mathbf{E}_t) + (\omega^2 \varepsilon \mu - \beta^2)\mathbf{E}_t + \nabla[\varepsilon^{-1} \nabla \cdot (\varepsilon \mathbf{E}_t)] = 0, \tag{5.7}$$

$$-\varepsilon \nabla \times (\varepsilon^{-1} \nabla \times \mathbf{H}_t) + (\omega^2 \varepsilon \mu - \beta^2)\mathbf{H}_t + \nabla[\mu^{-1} \nabla \cdot (\mu \mathbf{H}_t)] = 0, \tag{5.8}$$

$$\frac{(\omega^2 \varepsilon \mu - \beta^2)}{\varepsilon} \nabla \cdot \left[\frac{\varepsilon \nabla E_z}{(\omega^2 \varepsilon \mu - \beta^2)} \right] + (\omega^2 \varepsilon \mu - \beta^2)E_z$$
$$- \nabla \left[\ln \frac{\varepsilon \mu}{(\omega^2 \varepsilon \mu - \beta^2)} \right] \cdot \left[\mathbf{k} \times \frac{\omega \mu}{\beta} \nabla H_z \right] = 0, \tag{5.9}$$

$$\frac{(\omega^2 \varepsilon \mu - \beta^2)}{\mu} \nabla \cdot \left[\frac{\mu \nabla H_z}{(\omega^2 \varepsilon \mu - \beta^2)} \right] + (\omega^2 \varepsilon \mu - \beta^2)H_z$$
$$+ \nabla \left[\ln \frac{\varepsilon \mu}{(\omega^2 \varepsilon \mu - \beta^2)} \right] \cdot \left[\mathbf{k} \times \frac{\omega \varepsilon}{\beta} \nabla E_z \right] = 0. \tag{5.10}$$

On the other hand, as shown also in Appendix 5A.1, the transverse and axial components of electric and magnetic fields are related as

$$\mathbf{E}_t = -j \frac{\beta}{(\omega^2 \varepsilon \mu - \beta^2)} \left[\nabla E_z - \frac{\omega \mu}{\beta} \mathbf{k} \times \nabla H_z \right], \tag{5.11}$$

$$\mathbf{H}_t = -j \frac{\beta}{(\omega^2 \varepsilon \mu - \beta^2)} \left[\nabla H_z + \frac{\omega \varepsilon}{\beta} \mathbf{k} \times \nabla E_z \right]. \tag{5.12}$$

Note that in these six equations both ε and μ are assumed to be spatially varying quantities.

5.2.2   Vectorial Wave and Scalar Wave Equations

The refractive index n of a material is $n = (\varepsilon\mu/\varepsilon_0\mu_0)^{1/2}$, where ε_0 and μ_0 denote the vacuum permittivity and permeability, respectively. Since we can consider that $\mu = \mu_0$ in actual fiber materials, the spatial variation of n stems from that of ε. Therefore, only ε is given as a function of position as ε (x, y, z) in Eqs. (5.7) and (5.8), and these can be rewritten, after some computations using vector identities, as

$$\nabla^2 \mathbf{E}_t + (\omega^2\varepsilon\mu_0 - \beta^2)\mathbf{E}_t + \nabla\left[\frac{\nabla\varepsilon}{\varepsilon} \cdot \mathbf{E}_t\right] = 0, \tag{5.13}$$

$$\nabla^2 \mathbf{H}_t + (\omega^2\varepsilon\mu_0 - \beta^2)\mathbf{H}_t + \frac{\nabla\varepsilon}{\varepsilon} \times (\nabla \times \mathbf{H}_t) = 0. \tag{5.14}$$

These equations comprise two transverse field components $[E_x$ and E_y or E_r and E_θ for Eq. (5.13), and H_x and H_y or H_r and H_θ for Eq. (5.14)] simultaneously because of the presence of the third terms. (If the third terms are absent as in uniform-core fibers, two transverse field components will be separable.) In the theory of optical fibers, such an equation written in terms of vectorial variables (i.e., an equation that cannot be separated into scalar components) is called a vectorial wave equation. Generally, wave equations describing wave phenomena in nonuniform-core fibers are always vectorial wave equations when no approximation is made.

However, when $(\nabla\varepsilon/\varepsilon)$ is small enough, we may neglect the third terms in Eqs. (5.13) and (5.14) and write

$$\nabla^2 \mathbf{E}_t + [\omega^2\varepsilon(r)\mu_0 - \beta^2]\mathbf{E}_t = 0, \tag{5.15}$$

$$\nabla^2 \mathbf{H}_t + [\omega^2\varepsilon(r)\mu_0 - \beta^2]\mathbf{H}_t = 0. \tag{5.16}$$

The assumption that $\nabla\varepsilon/\varepsilon$ is small includes the "small index gradient" and/or "weakly guiding" approximations. We may further separate the above two equations into the x and y components to write

$$\nabla^2 E_a + [\omega^2\varepsilon(r)\mu_0 - \beta^2]E_a = 0 \quad (E_a = E_x, E_y), \tag{5.17}$$

$$\nabla^2 H_a + [\omega^2\varepsilon(r)\mu_0 - \beta^2]H_a = 0 \quad (H_a = H_x, H_y). \tag{5.18}$$

In wave theory these equations are called scalar wave equations.

5.2.3   Scalar Wave Equation in Cylindrical Coordinates

We now derive the scalar wave equation in cylindrical coordinates from Eqs. (5.17) and (5.18). In the following we consider only the electric field, i.e., Eq. (5.17).

Equation (5.17) can be rewritten in terms of r and θ as (Appendix 4A.1)

$$\nabla^2 E_r - \frac{1}{r^2} E_r - \frac{2}{r^2}\frac{\partial E_\theta}{\partial \theta} + [\omega^2\varepsilon(r)\mu_0 - \beta^2]E_r = 0, \tag{5.19}$$

$$\nabla^2 E_\theta - \frac{1}{r^2} E_\theta + \frac{2}{r^2}\frac{\partial E_r}{\partial \theta} + [\omega^2\varepsilon(r)\mu_0 - \beta^2]E_\theta = 0. \tag{5.20}$$

In these equations, E_r and E_θ are not separable. However, if we write (following Kurtz and Streifer [1]) E_r and E_θ as

$$E_r^{(i)} = \pm jR^{(i)}(r)e^{-jn\theta} \tag{5.21}$$
$$E_\theta^{(i)} = R^{(i)}(r)e^{-jn\theta} \qquad (i = 1, 2), \tag{5.22}$$

the new variable $R^{(i)}$ is given as the solution of

$$\frac{1}{r}\frac{d}{dr}\left(r\frac{dR^{(i)}}{dr}\right) + \left[\omega^2\varepsilon(r)\mu_0 - \beta^2 - \frac{(n \mp 1)^2}{r^2}\right]R^{(i)} = 0 \qquad (i = 1, 2). \tag{5.23}$$

In Eqs. (5.21)–(5.23), $i = 1$ and $i = 2$ represent two linearly independent solutions, and the upper and lower signs correspond to $i = 1$ and $i = 2$, respectively. The derivation of Eq. (5.23) is given in Appendix 5A.2.

Equation (5.23) is the scalar wave equation in cylindrical coordinates. More details of the mathematical implication of this equation will be shown in the discussion for each mode.

For these discussions, an expression similar to, but slightly different from, Eq. (5.23) will be derived. If we write

$$E_x \quad (\text{or } E_y) = R(r)e^{-jm\theta} \tag{5.24}$$

and substitute this into Eq. (5.17), we obtain directly

$$\frac{1}{r}\frac{d}{dr}\left(r\frac{dR}{dr}\right) + \left[\omega^2\varepsilon(r)\mu_0 - \beta^2 - \frac{m^2}{r^2}\right]R = 0. \tag{5.25}$$

This equation will be cited later together with Eq. (5.23) in Section 5.2.5.

5.2.4   Wave Equations and Their Solutions for Each Mode

To reveal clearly the approximations and assumptions implied in the analysis, derivations of the wave equation for each mode are required as seen in the succeeding sections.

We start from Eqs. (5.7)–(5.12). However, in the following analysis, we put Eqs. (5.7) and (5.8) aside; we first compute E_z and H_z from the differential equations (5.9) and (5.10) and then compute the transverse field components using Eqs. (5.11) and (5.12). We make the following assumptions and definition.

(1)   The permittivity ε is a function of only the radial coordinate r and is

$$\varepsilon(r) = \varepsilon_1[1 - f(r)], \tag{5.26}$$

where ε_1 denotes the maximum value of ε in the core; hence $f(r) \geq 0$.

(2)   The fiber material is nonmagnetic; i.e.,

$$\mu = \mu_0. \tag{5.27}$$

(3)   The following dimensionless parameter representing the phase constant is defined and used:

$$\chi = 1 - \beta^2/\omega^2\varepsilon_1\mu_0. \tag{5.28}$$

Let us express E_z and H_z as products of functions of r and θ as

$$E_z = (\omega^2\varepsilon_1\mu_0/\beta)\Phi(r)\cos(n\theta + \varphi_n), \tag{5.29}$$

$$H_z = \omega\varepsilon_1\Psi(r)\sin(n\theta + \varphi_n), \tag{5.30}$$

where n denotes an integer, $\varphi_n = 0$ or $\pi/2$, and $\Phi(r)$ and $\Psi(r)$ are dimensionless functions. Substituting these equations into Eqs. (5.9) and (5.10), we obtain

$$\left(\frac{\chi - f}{1 - f}\right)\frac{1}{r}\frac{d}{dr}\left[\left(\frac{1 - f}{\chi - f}\right)r\frac{d\Phi}{dr}\right] + \left[\omega^2\varepsilon_1\mu_0(\chi - f) - \frac{n^2}{r^2}\right]\Phi$$

$$+ \frac{n}{r}\Psi\left(\frac{\chi - f}{1 - f}\right)\frac{d}{dr}\left(\frac{1 - f}{\chi - f}\right) = 0, \tag{5.31}$$

$$(\chi - f)\frac{1}{r}\frac{d}{dr}\left[\frac{1}{\chi - f}r\frac{d\Psi}{dr}\right] + \left[\omega^2\varepsilon_1\mu_0(\chi - f) - \frac{n^2}{r^2}\right]\Psi$$

$$+ \frac{n}{r}\Phi(\chi - f)\frac{d}{dr}\left(\frac{1}{\chi - f}\right) = 0. \tag{5.32}$$

On the other hand, if we put Eqs. (5.29) and (5.30) into (5.11) and (5.12), we may express the transverse field components in terms of Φ and Ψ as

$$E_r = -j \frac{1}{\chi - f} \left[\frac{d\Phi}{dr} + \frac{n}{r} \Psi \right] \cos(n\theta + \varphi_n), \tag{5.33}$$

$$E_\theta = j \frac{1}{\chi - f} \left[\frac{d\Psi}{dr} + \frac{n}{r} \Phi \right] \sin(n\theta + \varphi_n), \tag{5.34}$$

$$H_r = -j \frac{\beta}{\omega\mu_0} \frac{1}{\chi - f} \left[\frac{d\Psi}{dr} + \left(\frac{1-f}{1-\chi} \right) \frac{n}{r} \Phi \right] \sin(n\theta + \varphi_n), \tag{5.35}$$

$$H_\theta = -j \frac{\beta}{\omega\mu_0} \frac{1}{\chi - f} \left[\left(\frac{1-f}{1-\chi} \right) \frac{d\Phi}{dr} + \frac{n}{r} \Psi \right] \cos(n\theta + \varphi_n). \tag{5.36}$$

The computations for obtaining Eqs. (5.31)–(5.36) are rather tedious but essentially are only simple algebraic manipulations.

At first glance, Eqs. (5.31)–(5.36) might look complicated. However, if we group the solution into four cases, the equations become much simpler. Those four solutions correspond to the TE, TM, EH, and HE modes mentioned in Sections 4.3.2 and 4.3.3.

We first assume that $n = 0$ in Eqs. (5.31) and (5.32); i.e., the solution is axially symmetric. Then Eqs. (5.31) and (5.32), which originally were simultaneous equations, are separated into two independent differential equations in terms of Φ and Ψ. Therefore, we consider for simplicity two independent cases:

(1) when $\Phi = 0$, $\Psi \neq 0$, and
(2) when $\Phi \neq 0$, $\Psi = 0$.

In the following (Sections A and B), it is revealed that these two cases correspond to TE modes ($E_z = 0$) and TM modes ($H_z = 0$), respectively. As far as $n = 0$, any solution of the equation should be expressed as a linear combination of such TE and TM modes. Cases in which $n \neq 0$ correspond to the hybrid modes and are described in Section C.

A. TE Modes

We consider cases when $n = 0$ and $\Phi = 0$. Then Eqs. (5.29), (5.33), and (5.36) lead directly to

$$E_z = E_r = H_\theta = 0. \tag{5.37}$$

Such cases obviously correspond to TE modes because $E_z = 0$. If we write

$$R(r) = j \frac{1}{\chi - f} \frac{d\Psi}{dr} \tag{5.38}$$

and substitute this into Eq. (5.32), we obtain

$$\Psi = j\frac{1}{\omega^2\varepsilon_1\mu_0}\left[\frac{dR}{dr} + \frac{1}{r}R\right]. \tag{5.39}$$

Further, substituting this equation into Eq. (5.38), we have the differential equation

$$\frac{1}{r}\frac{d}{dr}\left(r\frac{dR}{dr}\right) + \left[\omega^2\varepsilon_1\mu_0(\chi - f) - \frac{1}{r^2}\right]R = 0 \tag{5.40}$$

which is the basic wave equation for TE modes. On using $R(r)$ obtained by solving the preceding equation, we can express, from Eqs. (5.30), (5.34), (5.35), (5.38), and (5.39), the nonzero components of the electromagnetic field as

$$E_\theta = R(r), \tag{5.41}$$

$$H_r = -(\beta/\omega\mu_0)R(r), \tag{5.42}$$

$$H_z = j(1/\omega\mu_0)[dR/dr + r^{-1}R]. \tag{5.43}$$

B. TM *Modes*

Next we consider cases when $n = 0$ and $\Psi = 0$. In such cases Eqs. (5.30), (5.34), and (5.35) lead directly to

$$H_z = E_\theta = H_r = 0. \tag{5.44}$$

Such cases correspond to TM modes because $H_z = 0$. If we then write

$$R(r) = -j\frac{1}{\chi - f}\frac{d\Phi}{dr} \tag{5.45}$$

and substitute this into Eq. (5.31), we obtain

$$\Phi = -j\frac{1}{\omega^2\varepsilon_1\mu_0}\left[\frac{dR}{dr} + \frac{1}{r}R\right]. \tag{5.46}$$

Further, substituting this equation into Eq. (5.45), we obtain the differential equation

$$\frac{1}{r}\frac{d}{dr}\left(r\frac{dR}{dr}\right) + \left[\omega^2\varepsilon_1\mu_0(\chi - f) - \frac{1}{r^2}\right]R = 0. \tag{5.47}$$

In deriving Eqs. (5.46) and (5.47), an approximation $|rf'(r)| \ll 1$ has been made, where the prime denotes a differentiation. Note that such an assumption is not required when the TE mode is considered. This assumption corresponds, of course, to that made in deriving Eqs. (5.15) and (5.16), i.e., $(\Delta\varepsilon/\varepsilon)$ is small enough.

Equation (5.47) is the basic wave equation for TM modes. Using the solution $R(r)$ of this equation, the nonzero field components are, from (5.29), (5.33), (5.36), (5.45), and (5.46),

$$E_r = R(r), \tag{5.48}$$

$$E_z = -j\frac{1}{\beta}\left[\frac{dR}{dr} + \frac{1}{r}R\right], \tag{5.49}$$

$$H_\theta = \frac{\beta}{\omega\mu_0}R(r). \tag{5.50}$$

C. Hybrid Modes (EH and HE Modes)

We next consider cases when $n \neq 0$. For these cases we again introduce the following assumptions to facilitate the analysis, because otherwise the wave equation would remain vectorial.

We assume that the index difference between core and cladding is relatively small, so that the following approximations hold in Eqs. (5.31)–(5.36):

$$1 - f \simeq 1, \qquad 1 - \chi = \beta^2/\omega^2\varepsilon_1\mu_0 \simeq 1. \tag{5.51}$$

Apparently this seems to imply the weakly guiding assumption, but also includes the graded-core approximation because $1 - f$ is differentiated with respect to r in Eqs (5.31) and (5.32). As for the magnitude of $(\chi - f)$, no approximation is made because magnitudes of χ and f are comparable.

Under these approximations, Eqs. (5.31), (5.32), (5.35), and (5.36) are somewhat simplified to give

$$(\chi - f)\frac{1}{r}\frac{d}{dr}\left[\frac{r}{\chi - f}\frac{d\Phi}{dr}\right] + \left[\omega^2\varepsilon_1\mu_0(\chi - f) - \frac{n^2}{r^2}\right]\Phi$$
$$+ \frac{n}{r}(\chi - f)\Psi\frac{d}{dr}\left(\frac{1}{\chi - f}\right) = 0, \tag{5.52}$$

$$(\chi - f)\frac{1}{r}\frac{d}{dr}\left[\frac{r}{\chi - f}\frac{d\Psi}{dr}\right] + \left[\omega^2\varepsilon_1\mu_0(\chi - f) - \frac{n^2}{r^2}\right]\Psi$$
$$+ \frac{n}{r}(\chi - f)\Phi\frac{d}{dr}\left(\frac{1}{\chi - f}\right) = 0, \tag{5.53}$$

$$H_r = -j\frac{\beta}{\omega\mu_0}\frac{1}{\chi - f}\left[\frac{d\Psi}{dr} + \frac{n}{r}\Phi\right]\sin(n\theta + \varphi_n), \tag{5.54}$$

$$H_\theta = -j\frac{\beta}{\omega\mu_0}\frac{1}{\chi - f}\left[\frac{d\Phi}{dr} + \frac{n}{r}\Psi\right]\cos(n\theta + \varphi_n). \tag{5.55}$$

Note that Eqs. (5.33) and (5.34) remain unchanged. Next we introduce two new variables

$$\phi = (\Phi + \Psi)/2, \tag{5.56}$$

$$\psi = (\Phi - \Psi)/2. \tag{5.57}$$

If we take the sum and difference of Eqs. (5.52) and (5.53) and rewrite these equations in terms of ϕ and ψ, we can obtain

$$(\chi - f)\frac{1}{r}\frac{d}{dr}\left[\frac{r}{\chi - f}\frac{d\psi}{dr}\right] + \left[\omega^2 \varepsilon_1 \mu_0 (\chi - f) - \frac{n^2}{r^2}\right]\psi$$

$$- \frac{n}{r}(\chi - f)\psi \frac{d}{dr}\left(\frac{1}{\chi - f}\right) = 0, \tag{5.58}$$

$$(\chi - f)\frac{1}{r}\frac{d}{dr}\left[\frac{r}{\chi - f}\frac{d\phi}{dr}\right] + \left[\omega^2 \varepsilon_1 \mu_0 (\chi - f) - \frac{n^2}{r^2}\right]\phi$$

$$+ \frac{n}{r}(\chi - f)\phi \frac{d}{dr}\left(\frac{1}{\chi - f}\right) = 0. \tag{5.59}$$

It is now understood that if we make appropriate approximations [Eq. (5.51)] and transformation of the variables [Eqs. (5.56) and (5.57)], the original simultaneous differential equations [Eqs. (5.31) and (5.32)] which essentially correspond to a vectorial wave equation, can be separated into two independent equations for two scalar quantities ϕ and ψ, which are scalar wave equations. Hence, we may postulate that, as in the cases for TE and TM modes, we have two independent cases:

(1)   when $\phi = 0$, $\psi \neq 0$, and
(2)   when $\phi \neq 0$, $\psi = 0$.

Waves satisfying condition (1) are called EH modes, and those satisfying condition (2) are called HE modes. As shown in the following, both modes are the hybrid modes in which $E_z \neq 0$ and $H_z \neq 0$.

EH Modes   Since $n \neq 0$ is assumed originally and n is positive, $n \geq 1$ in this case. From $\phi = 0$ we obtain $\psi = \Phi = -\Psi$. Hence, if we write

$$R(r) = -j\frac{1}{\chi - f}\left[\frac{d\psi}{dr} - \frac{n}{r}\psi\right] \tag{5.60}$$

and substitute this into Eq. (5.58), we obtain

$$\psi = -j\frac{1}{\omega^2 \varepsilon_1 \mu_0}\left[\frac{dR}{dr} + \frac{n+1}{r}R\right]. \tag{5.61}$$

Furthermore, by substituting Eq. (5.61) into (5.60), we obtain the differential equation

$$\frac{1}{r}\frac{d}{dr}\left(r\frac{dR}{dr}\right) + \left[\omega^2\varepsilon_1\mu_0(\chi - f) - \frac{(n+1)^2}{r^2}\right]R = 0. \qquad (5.62)$$

This equation is the basic wave equation for EH modes. If we obtain $R(r)$ by solving this equation, we may express the field components, from Eqs. (5.29), (5.30), (5.33)–(5.36), (5.60), and (5.61), as

$$E_r = R(r)\cos(n\theta + \varphi_n), \qquad (5.63)$$

$$E_\theta = R(r)\sin(n\theta + \varphi_n), \qquad (5.64)$$

$$E_z = -j\frac{1}{\beta}\left[\frac{dR}{dr} + \frac{n+1}{r}R\right]\cos(n\theta + \varphi_n), \qquad (5.65)$$

$$H_r = -\frac{\beta}{\omega\mu_0}R(r)\sin(n\theta + \varphi_n), \qquad (5.66)$$

$$H_\theta = \frac{\beta}{\omega\mu_0}R(r)\cos(n\theta + \varphi_n), \qquad (5.67)$$

$$H_z = j\frac{1}{\omega\mu_0}\left[\frac{dR}{dr} + \frac{n+1}{r}R\right]\sin(n\theta + \varphi_n). \qquad (5.68)$$

HE Modes   In this case we may again assume $n \geq 1$. Since $\psi = 0$, we can write $\phi = \Phi = \Psi$. Therefore, if we write

$$R(r) = -j\frac{1}{\chi - f}\left[\frac{d\phi}{dr} + \frac{n}{r}\phi\right] \qquad (5.69)$$

and subsitute this into Eq. (5.59), we obtain

$$\phi = -j\frac{1}{\omega^2\varepsilon_1\mu_0}\left[\frac{dR}{dr} - \frac{n-1}{r}R\right]. \qquad (5.70)$$

By substituting Eq. (5.70) into (5.69), we obtain the differential equation

$$\frac{1}{r}\frac{d}{dr}\left(r\frac{dR}{dr}\right) + \left[\omega^2\varepsilon_1\mu_0(\chi - f) - \frac{(n-1)^2}{r^2}\right]R = 0 \qquad (5.71)$$

which is the basic wave equation for HE modes. If we obtain $R(r)$ by solving this equation, we may express the field components, from Eqs. (5.29), (5.30),

(5.33)–(5.36), (5.69), and (5.70), as

$$E_r = R(r)\cos(n\theta + \varphi_n), \tag{5.72}$$

$$E_\theta = -R(r)\sin(n\theta + \varphi_n), \tag{5.73}$$

$$E_z = -j\frac{1}{\beta}\left[\frac{dR}{dr} - \frac{n-1}{r}R\right]\cos(n\theta + \varphi_n), \tag{5.74}$$

$$H_r = \frac{\beta}{\omega\mu_0}R(r)\sin(n\theta + \varphi_n), \tag{5.75}$$

$$H_\theta = \frac{\beta}{\omega\mu_0}R(r)\cos(n\theta + \varphi_n), \tag{5.76}$$

$$H_z = -j\frac{1}{\omega\mu_0}\left[\frac{dR}{dr} - \frac{n-1}{r}R\right]\sin(n\theta + \varphi_n). \tag{5.77}$$

5.2.5 Unified Scalar Wave Equation and LP-Mode Concept

Comparing Eqs. (5.40), (5.47), (5.62), and (5.71), we now find that the first step of the analysis of the electromagnetic field in a nonuniform-core fiber is to solve a differential equation of the form identical to Eq. (5.25):

$$\frac{1}{r}\frac{d}{dr}\left(r\frac{dR}{dr}\right) + \left[\omega^2\varepsilon(r)\mu_0 - \beta^2 - \frac{m^2}{r^2}\right]R = 0, \tag{5.78}$$

in which the parameter m is given by

$$m = \begin{cases} 1 & \text{for TE and TM modes } (n = 0), \tag{5.79a} \\ n+1 & \text{for EH modes } (n \geq 1), \tag{5.79b} \\ n-1 & \text{for HE modes } (n \geq 1). \tag{5.79c} \end{cases}$$

The implications of the parameters m and n have already been considered in Sections 4.3.5 and 4.3.8. Equation (5.78) is the unified scalar wave equation for nonuniform-core optical fibers.

We have now thoroughly investigated for each mode the physical implications of, and approximations needed to derive, a unified scalar wave equation such as Eqs. (5.25) and (5.78). For TE modes, Eq. (5.78) can be derived without any approximations. For TM modes, the assumption of a small index gradient ($|rf'(r)| \ll 1$) is needed. For EH and HE modes, in addition to the assumption of a small index gradient, the weakly guiding approximation ($f \ll 1$) is required.

Equations (5.78) and (5.79) show that various modes can be analyzed in a unified manner based on the parameter m which is the rotational mode number related to the Cartesian coordinates and is used in LP-mode designations [see Eq. (5.24) and Section 4.3.8]. This means that the LP-mode concept is effective even in nonuniform-core fibers, provided that the assumption of small index gradient and the weakly guiding approximation are made.

5.2.6   Relation between Axial and Transverse Field Functions

We should clarify here the relation between the wave equation obtained for uniform-core fibers [Eq. (4.25)] and the scalar wave equation for nonuniform-core fibers [Eq. (5.78)]. In the theory of uniform-core fibers, the axial field function R_z is used [Eq. (4.25)], whereas in the theory of nonuniform-core fibers, transverse field function R is used.

Such two different functions have been used merely for the sake of simplicity of the formulations. We can directly relate R_z and R by using the expressions in Section 5.2.4. We first assume for the comparison of the two functions that the core is now uniform, and note that R_z used in Eq. (4.25) can stand for both E_z and H_z [see Eq. (4.23)]. Then from Eqs. (5.43), (5.49), (5.65), (5.68), (5.74), and (5.77), we can write

$$R_z \propto \frac{dR}{dr} + \frac{1}{r} R \qquad \text{for TE and TM modes,} \qquad (5.80)$$

$$R_z \propto \frac{dR}{dr} + \frac{n+1}{r} R \qquad \text{for EH modes,} \qquad (5.81)$$

$$R_z \propto \frac{dR}{dr} - \frac{n-1}{r} R \qquad \text{for HE modes.} \qquad (5.82)$$

Cases of TE, TM, and EH modes can immediately be addressed in a unified manner. If we use m defined as in Eq. (5.79), we may unify Eqs. (5.80) and (5.81) to obtain

$$R_z \propto (dR/dr) + (m/r)R. \qquad (5.83)$$

On the other hand, the solution of Eq. (5.78) in a uniform core is given in the form (see Section 4.3.1)

$$R \propto J_m(ur/a). \qquad (5.84)$$

Substituting Eq. (5.84) into (5.83) and using Bessel-function formulas given in Appendix 4A.2, we obtain

$$R_z \propto J_{m-1}(ur/a). \tag{5.85}$$

For HE modes, Eq. (5.82) leads to

$$R_z \propto (dR/dr) - (m/r)R. \tag{5.86}$$

Since Eq. (5.84) also holds for HE modes, substituting it into Eq. (5.86) and again after some computations using the Bessel-function formulas, we obtain

$$R_z \propto J_{m+1}(ur/a). \tag{5.87}$$

Equations (5.85) and (5.87) can be unified to give

$$R_z \propto J_n(ur/a), \tag{5.88}$$

which is equal to the solution obtained in Chapter 4 [Eq. (4.28)]. The same relations can be derived for the field in the cladding region.

5.2.7   Boundary Conditions

To compute the propagation characteristics of nonuniform-core fibers, we should solve Eq. (5.78). In most practical fibers, the refractive index varies in the core but is constant in the cladding; moreover, an index step is often present at the core–cladding boundary. Therefore, it is usually much more convenient to solve the wave equation in the core and cladding regions separately and connect those solutions at the core–cladding boundary in accordance with the physical boundary conditions.

Exact boundary conditions are given in Eqs. (4.51)–(4.56). However, under the weakly guiding approximation, i.e., when the permittivity change at the core–cladding boundary is small, all these equations can be approximated to give only two equations. If the transverse field functions in the core and cladding are denoted by $R(r)$ and $R_{clad}(r)$, respectively, those approximate boundary conditions are

$$R(a) = R_{clad}(a), \tag{5.89}$$

$$[dR/dr]_{r=a} = [dR_{clad}/dr]_{r=a}. \tag{5.90}$$

Note that all of the continuity conditions for the tangential and normal field components can be included in these two equations provided $[\varepsilon(a+0) - \varepsilon(a-0)]/\varepsilon(a-0) \ll 1$.

The function $R_{clad}(r)$ satisfies, as well as in uniform-core fibers; a differential equation having a form identical to Eq. (5.78) with $\varepsilon(r)$ being replaced by ε_2(const). As described in Section 4.3.1, the solution of this equation is given as a modified Bessel function in terms of the normalized radial coordinate.

5.2.8 Scalar Wave Analysis

Analysis of propagation characteristics based on the scalar wave equation [Eq. (5.78)] is called a scalar wave analysis. (Such an approximate analysis is sometimes called a TEM approximation; however, this term is inappropriate because axial field components can also be derived by this approach.)

Known methods for solving Eq. (5.78) are the (1) WKB method, (2) Rayleigh–Ritz method, (3) power-series expansion method, (4) finite element method, and (5) staircase-approximation method. Among these, the WKB method has an advantage in that a physical picture is relatively clear. However, the conventional WKB analysis is useful only for thick fibers in which many modes can propagate. For those fibers in which only a few modes propagate, the error of the WKB method intolerably increases. This method is not applicable to single-mode fibers.*

In the following, the WKB method, the Rayleigh–Ritz method and the power-series expansion method are described as examples of scalar wave analysis. The finite element method and the staircase-approximation method will be presented as examples of vectorial wave analysis.

5.2.9 Vectorial Wave Analysis

An analysis based on the vectorial wave equation [Eqs. (5.7)–(5.12)] is called vectorial wave analysis.

Known methods of vectorial wave analysis are the (1) finite element method, (2) staircase-approximation method, (3) direct integration method, and (4) perturbation method. In the perturbation method, the solution of the scalar wave equation is obtained first, and it is then put into the vectorial wave equation to compute the correction (perturbation) terms. This method is often useful for evaluating the effect of a deviation from a given (e.g., quardratic) index profile; however, this is not a universal method. The finite element method and the staircase-approximation method will be described as examples of vectorial wave analysis.

---

* Recent works reveal that we can improve the conventional WKB analysis to make it applicable to the close-to-cutoff modes and useful for the analysis of relatively thin or even single-mode fibers [2].

5.3   Analysis of Nonuniform-Core Fibers by the WKB Method

5.3.1   Basic Concepts

If we write*

$$\hat{R}(r) = \sqrt{r}\,R(r) \tag{5.91}$$

and substitute this in the scalar wave equation for nonuniform-core fibers [Eq. (5.78)], we obtain

$$\frac{d^2\hat{R}}{dr^2} + [E - U(r)]\hat{R} = 0, \tag{5.92}$$

where the new parameters are defined as

$$E = k^2 n_1^2 - \beta^2, \tag{5.93}$$

$$U(r) = [k^2 n_1^2 - k^2 n^2(r)] + \frac{(m^2 - \frac{1}{4})}{r^2}. \tag{5.94}$$

In these equations, $k = \omega/c$ and n_1 is the maximum refractive index in the core. Note that $k^2 n_1^2$ is added both in Eqs. (5.93) and (5.94) merely to make E and $U(r)$ positive.

Equation (5.92) has a same form as Schroedinger's wave equation [5]. The nature of the solution of an equation of this type is well known:

(1)   In the range of r where $U(r) < E$, $\hat{R}(r)$ is given as an oscillatory function of r.

(2)   In the range of r where $U(r) > E$, $\hat{R}(r)$ is given as an exponentially varying function of r.

If we assume for example that the refractive-index profile is something like Fig. 5.1a, the function $U(r)$ will have a form as shown in upper graphs in Fig. 5.2. On the other hand, E is a constant determined by the propagation constant β of the mode under consideration.

The three upper graphs in Fig. 5.2 show typical relations of $U(r)$ and E. It is seen that the oscillatory solution of $\hat{R}(r)$ appears in the hatched regions, whose "height" corresponds to the spatial frequency of the oscillation; the higher the hatched region, the higher the spatial frequency. Outside the

---

* Two other methods are available to convert Eq. (5.78) to the form of Eq. (5.92): (1) to use a new variable x defined as $R(r) = \exp[x(r)]$ instead of R [3]; (2) to use a new variable y defined as $r = a \exp y$, where a denotes the core radius [4]. Method (1), however, is a somewhat approximate one.

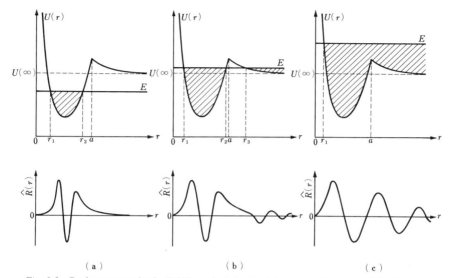

Fig. 5.2   Basic concepts in the WKB analysis for the (a) propagation mode, (b) leaky mode, and (c) radiation mode.

hatched regions, damped solutions are obtained. Therefore, the overall shape of each solution will be something like that shown in the lower graph.

Modes shown in Figs. 5.2a–5.2c are called the propagation mode, leaky mode, and radiation mode, respectively.

A. Propagation Mode

When the parameter E satisfies

$$0 < E < U(\infty), \tag{5.95}$$

i.e., when the propagation constant β lies in the range

$$kn_2 < \beta < kn_1, \tag{5.96}$$

the electromagnetic field decays in the outward radial direction as shown in Fig. 5.2a, so that the energy is predominantly confined in the core and propagates along the axis. Such an aspect of the optical energy transmission is called the trapped mode or the propagation mode.

A comment should be added concerning the field at the center of the core. Here the field is also expressed by an exponential-like function, but is not strictly exponential because $[E - U(r)]$ is a function of the position. Therefore, $R(r)$ $[= \hat{R}/\sqrt{r}]$ is not necessarily zero at $r = 0$.

B. Leaky Mode

When the parameters E and β satisfy

$$U(\infty) < E, \tag{5.97}$$

$$\beta < kn_2, \tag{5.98}$$

respectively, the electromagnetic field becomes spatially oscillatory in the cladding region as shown in Fig. 5.2b. This means that some electromagnetic energy may propagate outward. Hence the transmission power decays along the axial direction.

Such a mode whose propagation constant β is smaller than kn_2 is usually regarded to be beyond the cutoff condition (see Section 4.3.7). However, when E satisfies

$$U(\infty) < E < U(a) \tag{5.99}$$

(where a denotes the core radius) as is the case in Fig. 5.2b, the leakage of the electromagnetic energy can be rather small because of the "potential barrier" formed in the vicinity of the core–cladding boundary. The higher the potential barrier, the smaller the energy leakage. In such a transmission mode, although the solution in the cladding is radially oscillatory, its amplitude has to be rather small because the energy source (leakage) is originally small. Hence, the radial energy radiation and the axial transmission loss are both small. Such a transmission mode is called the "tunneling leaky mode" or simply the "leaky mode"; it behaves apparently like a "very lossy" propagation mode [6].

C. Radiation Mode

Those modes which are beyond cutoff and even satisfy

$$U(a) < E \tag{5.100}$$

as shown in Fig. 5.2c are called radiation modes. These are sometimes called refractive leaky modes because electromagnetic energy leaks not by tunneling but through refraction at the core–cladding boundary. The transmission loss of such modes is much higher than that in the leaky modes.

5.3.2 Analysis of Propagation Mode by the WKB Method

A. Basic Equations

To solve the space-dependent one-dimensional wave equation like Eq. (5.92) in an approximate manner, the WKB (Wentzel–Kramers–Brillouin) method has long been used in various fields including wave mechanics. This method is applied first to the propagation mode (Fig. 5.2a) [3, 4].

We first assume a refractive-index profile as shown in Fig. 5.1a and express it as

$$n^2(r) = n_1^2\{1 - f(r)\}, \tag{5.101}$$

$$0 \leqq f(r) \ll 1, \tag{5.102}$$

where n_1 is the refractive index on the axis. We rewrite Eqs. (5.92)–(5.94) for the oscillatory region where $E - U(r) > 0$:

$$\frac{d^2\hat{R}}{dr^2} + \beta_1^2 p(r)\hat{R} = 0, \tag{5.103}$$

$$p(r) = \frac{E - U(r)}{\beta_1^2} = 1 - \frac{\beta^2}{\beta_1^2} - f(r) - \frac{(m^2 - \frac{1}{4})}{\beta_1^2 r^2}, \tag{5.104}$$

$$\beta_1^2 = \omega^2 \varepsilon_0 \mu_0 n_1^2. \tag{5.105}$$

In these equations, β_1 denotes the propagation constant of a plane wave propagating in a medium with refractive index equal to n_1. The new variable $p(r)$ is positive everywhere in the oscillatory region.

For the damping region where $E - U < 0$, we rewrite Eqs. (5.92)–(5.94) using somewhat different notations as

$$\frac{d^2\hat{R}}{dr^2} - \beta_1^2 q(r)\hat{R} = 0, \tag{5.106}$$

$$q(r) = -1 + \frac{\beta^2}{\beta_1^2} + f(r) + \frac{(m^2 - \frac{1}{4})}{\beta_1^2 r^2}. \tag{5.107}$$

Note that $q(r)$ is also positive everywhere in the damping region.

B. Solution in Oscillatory Regions

We first assume that the solution in the oscillatory region ($r_1 < r < r_2$) can be written as

$$\hat{R}(r) = A \exp\{j\beta_1 s(r)\}. \tag{5.108}$$

Substituting this equation into Eq. (5.103), we obtain

$$j\beta_1^{-1} \frac{d^2 s}{dr^2} - \left(\frac{ds}{dr}\right)^2 + p(r) = 0. \tag{5.109}$$

Further, we expand $s(r)$ by the power of

$$\mu = \beta_1^{-1} \tag{5.110}$$

so that

$$s = s_0 + \mu s_1 + \mu^2 s_2 + \cdots . \tag{5.111}$$

Substituting this expression into Eq. (5.109), we obtain

$$0 = \{-s_0'^2 + p(r)\} + \mu(js_0'' - 2s_0's_1') + \mu^2(js_1'' - s_1'^2 - 2s_0's_2') + \cdots, \tag{5.112}$$

where a prime denotes differentiation with respect to r.

The parameter μ is proportional to the wavelength of light λ, because $\mu = \beta_1^{-1} = \lambda/2\pi n_1$. Therefore, the "short-wavelength approximation" requires use of only the lower-order terms in Eq. (5.111). In the ray theory described in Chapter 3, the extremity $\lambda \to 0$ has been considered. The ray theory corresponds to the case in which only the μ^0 term is considered, whereas the μ^1 term and further terms are neglected. In the WKB analyses, the μ^0 and μ^1 terms are considered

Taking the first two terms in Eq. (5.112), we obtain, from the condition that this equation holds regardless the value of μ,

$$-s_0'^2 + p(r) = 0, \tag{5.113}$$

$$js_0'' - 2s_0's_1' = 0. \tag{5.114}$$

These differential equations can be solved without difficulty; the solutions are

$$s_0(r) = \pm \int \sqrt{p(r)}\, dr + A_1, \tag{5.115}$$

$$s_1(r) = \tfrac{1}{2} j \ln \sqrt{p(r)} + A_2, \tag{5.116}$$

where A_1 and A_2 are unknown constants.

Using Eqs. (5.108), (5.111), (5.115), and (5.116), we obtain an approximate solution of Eq. (5.103) as

$$\hat{R}(r) = D\{p(r)\}^{-1/4} \sin\left[\beta_1 \int_{r_1}^{r} \sqrt{p(r)}\, dr + \varphi\right], \tag{5.117}$$

where D and φ are unknown constants.

C. Solution in Damping Regions

The equation to be solved is (5.106). After computations similar to those in the preceding case, we obtain the solution in the damping regions ($r < r_1$, $r_2 < r$),

$$\hat{R}(r) \propto \{q(r)\}^{-1/4} \exp\left[\pm \beta_1 \int \sqrt{q(r)}\, dr\right]. \tag{5.118}$$

If we define the constants and ranges of integral for two damping regions separately, we may write, considering the condition that the solutions do not diverge,

for $r < r_1$,

$$\hat{R}(r) = A\{q(r)\}^{-1/4} \exp\left[-\beta_1 \int_r^{r_1} \sqrt{q(r)}\,dr\right], \qquad (5.119)$$

for $r_2 < r$,

$$\hat{R}(r) = G\{q(r)\}^{-1/4} \exp\left[-\beta_1 \int_{r_2}^r \sqrt{q(r)}\,dr\right], \qquad (5.120)$$

where A and G are unknown constants.

5.3.3   Accuracy of WKB Solutions

Before proceeding to the next step of the analysis, we investigate the accuracy of the WKB solutions obtained in Section 5.3.2 [4]. It is easily proved by substitution that a differential equation

$$\frac{d^2\hat{R}}{dr^2} + \left[\beta_1^2 p(r) - \frac{5}{16}\left(\frac{p'}{p}\right)^2 + \frac{1}{4}\left(\frac{p''}{p}\right)\right]\hat{R} = 0 \qquad (5.121)$$

has a solution exactly identical to Eq. (5.117). Therefore, if

$$\left|\frac{1}{4}\frac{p''}{p} - \frac{5}{16}\left(\frac{p'}{p}\right)^2\right| \ll \beta_1^2 p, \qquad (5.122)$$

Eq. (5.117) is a good approximate solution to Eq. (5.103). In other words, when the variation of $p(r)$ within one wavelength is not excessive, the WKB method gives a good approximate solution.

On the other hand, as seen in Eq. (5.122), the accuracy is very poor in the vicinity of positions where $p(r) = 0$, i.e., for example in Fig. 5.2a, at $r = r_1$ and $r = r_2$. Such positions are called "turning points" because in the ray model the ray turns back into the oscillatory region at these points due to refraction.

5.3.4   Solutions in the Vicinity of Turning Points

The foregoing solutions lose sense in the vicinity of the turning points; solutions of a different type must be derived. Those solutions must agree with Eqs. (5.117), (5.119), and (5.120) asymptotically. Mathematically, such solutions should exist and be obtained exactly because Eqs. (5.103) and

(5.106) are both regular at the turning points. However, only approximate solutions will be shown in the following.

We first consider the inner turning point: $r = r_1$. The damping and oscillatory regions are given as $r < r_1$ and $r_1 < r$, respectively. When $p(r)$ varies moderately, we may write approximately, in the vicinity of the turning point,

$$p(r) = k(r - r_1), \tag{5.123}$$

where k is a positive constant. Such a point is called a turning point of the first order. It is known that in such a case, the solution of Eq. (5.103) is,

for $r \lesssim r_1$,

$$\hat{R}(r) = \{q(r)\}^{-1/4} \xi'^{1/4} [B \, \mathrm{Ai}(\xi') + C \, \mathrm{Bi}(\xi')], \tag{5.124}$$

$$\xi' = \left[\tfrac{3}{2} \beta_1 \int_r^{r_1} \sqrt{q(r)} \, dr \right]^{2/3}, \tag{5.125}$$

for $r_1 \lesssim r$,

$$\hat{R}(r) = \{p(r)\}^{-1/4} \xi^{1/4} [B \, \mathrm{Ai}(-\xi) + C \, \mathrm{Bi}(-\xi)], \tag{5.126}$$

$$\xi = \left[\tfrac{3}{2} \beta_1 \int_{r_1}^r \sqrt{p(r)} \, dr \right]^{2/3}, \tag{5.127}$$

where B and C are unknown constants, and Ai and Bi are transcendental functions called Airy functions. Definitions and graphs of the Airy functions will be found in Appendix 5A.3 and in Abramowitz and Stegun [7].

Next we consider the outer turning point $r = r_2$. In this case the oscillatory and damping regions are given as $r < r_2$ and $r_2 < r$, respectively. The solution of Eq. (5.103) is

for $r \lesssim r_2$,

$$\hat{R}(r) = \{p(r)\}^{-1/4} \eta^{1/4} [E \, \mathrm{Ai}(-\eta) + F \, \mathrm{Bi}(-\eta)], \tag{5.128}$$

$$\eta = \left[\tfrac{3}{2} \beta_1 \int_r^{r_2} \sqrt{p(r)} \, dr \right]^{2/3}, \tag{5.129}$$

for $r_2 \lesssim r$,

$$\hat{R}(r) = \{q(r)\}^{-1/4} \eta'^{1/4} [E \, \mathrm{Ai}(\eta') + F \, \mathrm{Bi}(\eta')], \tag{5.130}$$

$$\eta' = \left[\tfrac{3}{2} \beta_1 \int_{r_2}^r \sqrt{q(r)} \, dr \right]^{2/3}, \tag{5.131}$$

where E and F are unknown constants.

5.3.5 Connection of Solutions and Derivation of Proper Equation

So far we have obtained seven solutions for each region:

for $r < r_1$,   Eq. (5.119) including A,
for $r \lesssim r_1$,   Eq. (5.124) including B and C,
for $r_1 \lesssim r$,   Eq. (5.126) including B and C,
for $r_1 < r < r_2$,   Eq. (5.117) including D and φ,
for $r \lesssim r_2$,   Eq. (5.128) including E and F,
for $r_2 \lesssim r$,   Eq. (5.130) including E and F,
for $r_2 < r$,   Eq. (5.120) including G.

However, the eight constants (A–G and φ) are left unknown. To determine them, we have to connect the seven solutions over the turning points using the asymptotic solutions of Airy functions.

It is known that, for $|z| \gg 1$, Airy functions are approximated as [7]

$$\text{Ai}(z) \simeq (1/2\sqrt{\pi})z^{-1/4}\exp(-\tfrac{2}{3}z^{3/2}), \tag{5.132}$$

$$\text{Bi}(z) \simeq (1/\sqrt{\pi})z^{-1/4}\exp(\tfrac{2}{3}z^{3/2}), \tag{5.133}$$

$$\text{Ai}(-z) \simeq (1/\sqrt{\pi})z^{-1/4}\sin(\tfrac{2}{3}z^{3/2} + \tfrac{1}{4}\pi), \tag{5.134}$$

$$\text{Bi}(-z) \simeq (1/\sqrt{\pi})z^{-1/4}\cos(\tfrac{2}{3}z^{3/2} + \tfrac{1}{4}\pi). \tag{5.135}$$

Substituting these approximate formulas into the solutions for the vicinity of turning points and equating those to the oscillatory and damping solutions on both sides of the turning points, we obtain the following relations between the unknown constants:

$$A = B/2\sqrt{\pi}, \qquad G = E/2\sqrt{\pi},$$
$$C = 0, \qquad F = 0, \qquad D = B/\sqrt{\pi}, \qquad \varphi = \tfrac{1}{4}\pi, \tag{5.136}$$

$$D\sin\left[\beta_1 \int_{r_1}^{r}\sqrt{p(r)}\,dr + \tfrac{1}{4}\pi\right] = (E/\sqrt{\pi})\sin\left[\beta_1 \int_{r}^{r_2}\sqrt{p(r)}\,dr + \tfrac{1}{4}\pi\right]. \tag{5.137}$$

The details of the derivation of these formulas are given in Appendix 5A.4.

The proper equation giving the propagation constant can be derived from the preceding relations. First, in order that Eq. (5.137) holds for any r, as shown in Appendix 5A.4, item (8),

$$D = \pm E/\sqrt{\pi}, \tag{5.138}$$

$$\beta_1 \int_{r_1}^{r_2}\sqrt{p(r)}\,dr = (l - \tfrac{1}{2})\pi, \tag{5.139}$$

where l is an arbitrary integer. Using Eq. (5.104), we can rewrite Eq. (5.139) as

$$\int_{r_1}^{r_2} \left[k^2 n^2(r) - \beta^2 - \frac{(m^2 - \frac{1}{4})}{r^2} \right]^{1/2} dr = (l - \tfrac{1}{2})\pi. \tag{5.140}$$

This is the proper equation giving the propagation constant of the LP_{ml} mode [3, 8].

5.3.6 Analysis of Leaky Modes by the WKB Method

Let us next analyze the characteristics of the leaky modes following the preceding process. Equation (5.99) indicates that the propagation constant β of a leaky mode satisfies

$$k^2 n^2(a) - \frac{(m^2 - \frac{1}{4})}{a^2} < \beta^2 < k^2 n_2^2. \tag{5.141}$$

Computing the solutions of $R(r)$ for each mode and connecting those solutions at three turning points r_1, r_2, and r_3 shown in Fig. 5.2b, we obtain the proper equation as

$$\tan\left[\beta_1 \int_{r_1}^{r_2} \sqrt{p(r)}\, dr + \tfrac{1}{2}\pi \right] = \tfrac{1}{4} j \exp\left[-2\beta_1 \int_{r_2}^{r_3} \sqrt{q(r)}\, dr \right]. \tag{5.142}$$

When the potential barrier between r_2 and r_3 is relatively high and thick so that

$$\exp\left[-2\beta_1 \int_{r_2}^{r_3} \sqrt{q(r)}\, dr \right] \ll 1, \tag{5.143}$$

we can make an approximation $\tan(x + l\pi) \simeq x$, and rewrite Eq. (5.142) approximately as

$$\beta_1 \int_{r_1}^{r_2} \sqrt{p(r)}\, dr = (l + \tfrac{1}{2})\pi - \tfrac{1}{4} j \exp\left[-2\beta_1 \int_{r_2}^{r_3} \sqrt{q(r)}\, dr \right], \tag{5.144}$$

or by using Eq. (5.104), as

$$\int_{r_1}^{r_2} \left[k^2 n^2(r) - \beta^2 - \frac{(m^2 - \frac{1}{4})}{r^2} \right]^{1/2} dr = \left(\begin{array}{c} \text{the right-hand} \\ \text{side of (5.144)} \end{array} \right). \tag{5.145}$$

This is nothing but the proper equation giving the propagation constant of a leaky LP_{ml} mode.

The attenuation constant of a leaky mode is easily derived from (5.145). If we denote the proper value of an equation

$$\beta_1 \int_{r_1}^{r_2} \sqrt{p(r)}\, dr = (l - \tfrac{1}{2})\pi \tag{5.146}$$

by β_0 and the solution of Eq. (5.144) by

$$\beta = \beta_0 + \Delta\beta, \tag{5.147}$$

then $\Delta\beta$ is approximately purely imaginary, and we can compute the attenuation constant α of the leaky mode as

$$\alpha = j\Delta\beta \simeq \beta_1 \exp\left[-2\beta_1 \int_{r_2}^{r_3} \sqrt{q(r)}\, dr \right] \bigg/ 4\beta_0 \int_{r_1}^{r_2} \frac{dr}{\sqrt{p(r)}}. \tag{5.148}$$

5.4 Computation of Propagation Characteristics of Multimode Fibers Based on WKB Analysis

We can compute various propagation characteristics (number of propagating modes, delay time, impulse response, etc.) of nonuniform-core multimode fibers on the basis of the WKB analysis described in the preceding section [8].

5.4.1 Number of Propagating Modes

If we assume that $m \gg 1$ and $l \gg 1$, we can approximate Eq. (5.140) as

$$\int_{r_1(m)}^{r_2(m)} \left[k^2 n^2(r) - \beta^2 - \frac{m^2}{r^2} \right]^{1/2} dr \simeq l\pi. \tag{5.149}$$

Here we should note that the upper and lower limits of the integral r_1, r_2 are both functions of the rotational mode number m. Equation (5.149) may be inaccurate for small m and l. However, in the following analysis, we assume that a large number of modes are propagating and consider that the contribution of such lower-order modes are negligible. Henceforce we proceed as if Eq. (5.149) holds for any set of m and l.

If the maximum radial mode number of a propagation mode with a rotational mode number m is denoted by $l_{max}(m)$, we may write, directly from Eq. (5.149),

$$l_{max}(m) = \frac{1}{\pi} \int_{r_1(m)}^{r_2(m)} \left[k^2 n^2(r) - \beta_{min}^2 - \frac{m^2}{r^2} \right]^{1/2} dr, \tag{5.150}$$

where β_{min} is the minimum β associated with that given m described in the following.

Figure 5.3a shows how the U–r curve varies as m changes. From Eq. (5.150) we see that $l_{max}(m)$ modes can propagate for each m, i.e., $l = 1$, $2, \ldots, l_{max}$ (see Fig. 5.3b). The larger the rotational mode number m, the

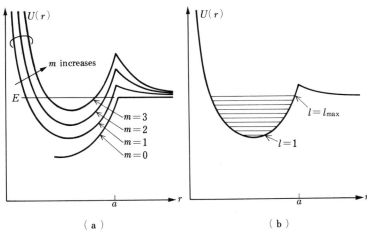

Fig. 5.3   Graphs depicting the calculation of the number of propagating modes. (a) $U(r)$-versus-r relations for various m. (b) Mode numbers l belonging to a specific m.

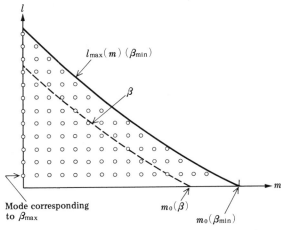

Fig. 5.4   Relation between the propagation constant β and mode numbers m, l.

smaller the maximum radial mode number $l_{\max}(m)$. This relation is illustrated in Fig. 5.4.

Next we notice that E shown in Fig. 5.2 decreases when β increases [see Eq. (5.93)], and hence in Figs. 5.3a and 5.3b, β becomes smaller for larger m (when l is constant) and for larger l (when m is constant). The maximum propagation constant β_{max} is given for $m = 0$, $l = 1$. The minimum propagation constant β_{min} is given, from the cutoff condition, by $\beta_{min} = kn_2$. Therefore, if the number of modes having a propagation constant between β and β_{max} is denoted by $v(\beta)$, it can be given approximately by counting the number of circles plotted beneath the lower left of the dashed curve in Fig. 5.4, as

$$v(\beta) = \frac{4}{\pi} \int_0^{m_0(\beta)} \int_{r_1(m)}^{r_2(m)} \left[k^2 n^2(r) - \beta^2 - \frac{m^2}{r^2} \right]^{1/2} dr\, dm. \qquad (5.151)$$

In this equation m_0 (β) is the maximum m for a given β (see Fig. 5.4). The factor 4 in the right-hand side of (5.151) has been given to express the fact that four conventional modes belong to most of each LP mode (refer to Tables 4.2 and 4.3). Note also that the summation for m is replaced by the integration with respect to m because $m_0 \gg 1$ is assumed.

From Eq. (5.151) we see that $[k^2 n^2 - \beta^2 - m^2/r^2]^{1/2}$ should be integrated in the hatched region of Fig. 5.5. Exchanging the order of integration, we can rewrite Eq. (5.151) as

$$v(\beta) = \frac{4}{\pi} \int_0^{r_2(0)} \int_0^{r\sqrt{k^2 n^2(r) - \beta^2}} \left[k^2 n^2(r) - \beta^2 - \frac{m^2}{r^2} \right]^{1/2} dm\, dr. \qquad (5.152)$$

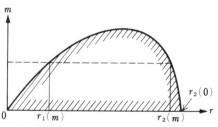

Fig. 5.5 Region of integral in Eq. (5.151).

In this expression, the upper limit of the integral m_0 (β) is replaced by

$$m_0 = r\sqrt{k^2 n^2(r) - \beta^2} \qquad (5.153)$$

because m_0 is obtained by putting $l = 0$ into Eq. (5.149). The integration with respect to m can be performed without difficulty to obtain

$$v(\beta) = \int_0^{r_2(0)} [k^2 n^2(r) - \beta^2] r\, dr. \qquad (5.154)$$

Note that the upper limit of the integration $r_2(0)$ is equal to the core diameter a, because $r_2(0)$ denotes r_2 for $m = 0$ (see Fig. 5.3a).

We consider here the so-called α-power refractive-index profile

$$n(r) = \begin{cases} n_1[1 - 2\Delta(r/a)^{\alpha}]^{1/2} & (0 \leq r \leq a), & (5.155a) \\ n_2 = n_1(1 - 2\Delta)^{1/2} & (a < r), & (5.155b) \end{cases}$$

where the power α is assumed to range between 1 and ∞. Profiles for various values of α are shown in Fig. 5.6. The usefulness of this profile lies in the fact that it approximates the actual profiles found in many fibers. For example, $\alpha = \infty$ corresponds to the uniform-core fiber.

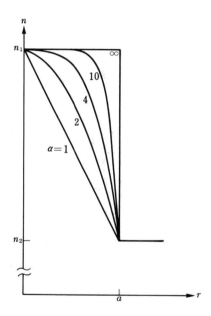

Fig. 5.6   The α-power refractive-index profile given by Eq. (5.155).

Substituting Eq. (5.155) into (5.154) and performing the integration, we obtain

$$v(\beta) = \alpha^2 k^2 n_1^2 \Delta \frac{\alpha}{\alpha + 2} \left(\frac{k^2 n_1^2 - \beta^2}{2k^2 n_1^2 \Delta} \right)^{(\alpha + 2)/\alpha} \tag{5.156}$$

The total number of propagating modes N is obtained by substituting $\beta = \beta_{\min} = kn_2$ into (5.156),

$$N = v(kn_2) = a^2 k^2 n_1^2 \Delta \frac{\alpha}{\alpha + 2}. \tag{5.157}$$

From Eq. (5.156), the propagation constant β is given by using N,

$$\beta = kn_1 \left[1 - 2\Delta \left(\frac{v}{N} \right)^{\alpha/(\alpha+2)} \right]^{1/2}. \tag{5.158}$$

This relation will be used in Section 5.4.2.

Before proceeding to the discussion on the dispersion relation, we derive some important relations concerning the number of modes. First, when $\alpha = \infty$ (uniform-core case), Eq. (5.158) is reduced to

$$\beta = kn_1[1 - 2\Delta(v/N)]^{1/2}. \tag{5.159}$$

Comparing this equation with Eq. (4.78), we find the simple relation

$$v/N = (u/v)^2 = x \tag{5.160}$$

for a uniform-core fiber, where x is the variable defined in Eq. (4.82). In other words, x is proportional to the number of LP modes having β greater than that corresponding to x.

Next, we derive a similar relation for an arbitrary α. Parameters u, v, and x were defined in Section 4.3 for uniform-core fibers. We now extend those definitations,

$$u^2 = (k^2 n_1^2 - \beta^2)a^2 \qquad \text{[see Eq. (4.43)]}, \tag{5.161}$$

$$v^2 = k^2 n_1^2 a^2 (2\Delta) \qquad \text{[see Eq. (4.76)]}, \tag{5.162}$$

$$x = u^2/v^2 \qquad \text{[see Eq. (4.82)]}, \tag{5.163}$$

to the nonuniform, α-power core fibers without any change. Then, from Eqs. (5.156) and (5.157), we obtain

$$v/N = (u/v)^{2(\alpha+2)/\alpha} = x^{(\alpha+2)/\alpha}. \tag{5.164}$$

5.4.2 Group Delay and Optimum Power α

As given in Eq. (4.95), the group delay (defined as the time required by an impulse to propagate over a unit length of fiber) is given by $(d\beta/d\omega)$. For α-power profiles, the group delay is obtained, by differentiating Eq. (5.158), considering that n_1 and Δ are both functions of ω, and hence, of k. The result of such a differentiation is

$$t = \frac{1}{c} \frac{d\beta}{dk} = \frac{N_1}{c} \left[1 + \Delta \frac{\alpha - 2 - y}{\alpha + 2} x + \frac{\Delta^2}{2} \frac{3\alpha - 2 - 2y}{\alpha + 2} x^2 + O(\Delta^3) \right]. \tag{5.165}$$

In this expression, N_1 is the group index defined by Eq. (4.98), which is approximately equal to n_1 [Eq. (4.99)], and y is the parameter defined by Olshansky [Eq. (4.100)].

Equation (5.165) indicates there exists an optimum value of α. The first term within the brackets gives a constant group delay. The variation of the group delay for various modes (multimode dispersion) is given in the second and third terms, where the mode number is represented by the variable x. Between these terms, the second term is predominant because $\Delta \ll 1$. Therefore, when

$$\alpha = 2 + y, \tag{5.166}$$

the dispersion will be minimized.

Olshansky's parameter y at $\lambda = 800$ nm, e.g., is about 0.3 when TiO_2 is doped to silica glass, as shown in Fig. 4.11d. According to latest data given by Presby and Kaminov [9], the value of y for silica glass is about 0.23 for the TiO_2 doping, 0.07 for GeO_2 doping, and $-0.23 - -0.33$ for B_2O_3 doping at $\lambda = 800$ nm.

5.4.3   Impulse Response

In the following, the impulse response of nonuniform-core fibers having α-power profile is computed. For simplicity it is assumed that $y = 0$.

An optical impulse having a delta-function waveform is launched into the fiber at the transmitting end. This impulse arrives at the receiving end after being widened by the multimode dispersion. Let the total power of the received impulse and the light power received between t and $t + dt$ be denoted by P and $p(t)\,dt$, respectively. The impulse response will then be express as

$$h(t)\,dt = p(t)\,dt/P. \tag{5.167}$$

If we assume that the total power is distributed equally to all modes, we may rewrite (5.167), replacing the power by the mode number, as

$$h(t) = N^{-1}\,dv(t)/dt, \tag{5.168}$$

where $v(t)$ denotes the number of modes arriving at the receiving end before time t.

In the following computations, the propagation constant is represented by x. The number of modes having propagation constant between x and $x + dx$ is expressed as

$$dv = [dv(x)/dx]\,dx. \tag{5.169}$$

The preceding two equations lead directly to

$$h(t) = N^{-1}(dv/dx)(dt/dx)^{-1} \qquad (5.170)$$

We first compute (dt/dx). Assuming that the fiber has a unit length and $y = 0$, and denoting the delay-time difference between the lowest mode and the mode under consideration by $\tau [= t(x) - t(0)]$, we obtain, from Eq. (5.165),

$$\tau \simeq \begin{cases} (N_1/c)\,\Delta[(\alpha - 2)/(\alpha + 2)]x & (\alpha \neq 2), & (5.171a) \\ (N_1/c)(\Delta^2/2)x^2 & (\alpha = 2). & (5.171b) \end{cases}$$

The delay-time difference between the highest mode $(x = 1)$ and the lowest mode $(x = 0)$ T is given, directly from the preceding equation, by

$$T = \begin{cases} (N_1/c)\,\Delta[(\alpha - 2)/(\alpha + 2)] & (\alpha \neq 2), & (5.172a) \\ (N_1/c)(\Delta^2/2) & (\alpha = 2). & (5.172b) \end{cases}$$

The quantity T is usually called the "delay difference between modes," or the "intermodal delay difference," or sometimes simply the "multimode dispersion." By using T, Eq. (5.171) is rewritten as

$$\tau = \begin{cases} Tx & (\alpha \neq 2), & (5.173a) \\ Tx^2 & (\alpha = 2). & (5.173b) \end{cases}$$

Note that $0 \leq x \leq 1$. From the foregoing equations we obtain (dt/dx),

$$\frac{dt}{dx} = \frac{d\tau}{dx} = \begin{cases} T & (\alpha \neq 2), & (5.174a) \\ 2Tx = 2\sqrt{T\tau} & (\alpha = 2). & (5.174b) \end{cases}$$

Next we compute (dv/dx). From Eq. (5.164),

$$dv/dx = [(\alpha + 2)/\alpha]Nx^{2/\alpha}. \qquad (5.175)$$

Expressing x in this equation by τ using Eq. (5.173), we obtain

$$\frac{dv}{dx} = \begin{cases} [(\alpha + 2)/\alpha]N(\tau/T)^{2/\alpha} & (\alpha \neq 2), & (5.176a) \\ 2N(\tau/T)^{1/2} & (\alpha = 2). & (5.176b) \end{cases}$$

Substituting Eqs. (5.174) and (5.176) into (5.170), we obtain

$$h(\tau) = \begin{cases} [(\alpha + 2)/\alpha](\tau^{2/\alpha}/T^{(\alpha + 2)/\alpha}) & (\alpha \neq 2), & (5.177a) \\ 1/T & (\alpha = 2), & (5.177b) \end{cases}$$

which give the impulse response for α-power nonuniform-core fibers.

The impulse responses given by Eq. (5.177) are shown in Fig. 5.7 [8]. For the uniform-core fiber $(\alpha = \infty)$, the impulse spread T is largest $(Tc/N_1 = \Delta,$

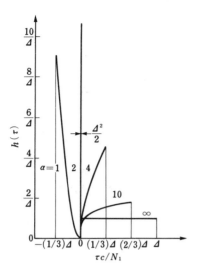

Fig. 5.7 Impulse response of α-power non-uniform-core fibers (after Gloge *et al.* [8]; Copyright 1973, American Telephone and Telegraph Company; Reprinted by permission).

hence $T = N_1\Delta/c$), and the waveform of the impulse response is flat-topped. As the power α decreases, the impulse response becomes narrower, until at $\alpha = 2$ where it becomes the narrowest ($Tc/N_1 = \Delta^2/2$, hence $T = N_1\Delta^2/2c$). When $\alpha < 2$, higher-order modes arrive earlier than the lowest mode ($x = 0$). At $\alpha = 1$ and $\alpha = 4$, the intermodal delay difference is one-third that of a uniform-core fiber.

Comparison of Eqs. (5.172a) and (5.172b) shows that the intermodal delay difference of a quadratic-core fiber ($\alpha = 2$) is ($\Delta/2$) times that of a uniform-core fiber. The factor ($\Delta/2$) is typically of the order of magnitude of 10^{-3}. This value suggests that the multimode dispersion could be reduced remarkably by shaping the refractive-index profile properly.

5.5   Analysis of Nonuniform-Core Fibers by Rayleigh–Ritz Method

In this section, analysis (computation of propagation constant and delay characteristics) of nonuniform-core fibers having arbitrary index profiles is performed by using Rayleigh–Ritz method. The scalar wave equation [Eq. (5.78)] is first translated into a variational problem under the given boundary conditions [Eqs. (5.89) and (5.90)]. The variational problem is then solved by using the Rayleigh–Ritz method, and the dispersion equation for an arbitrary nonuniform-core fiber is derived. Some parts of the following computations are rather tedious and are described in Appendixes 5A.5–5A.8.

In the succeeding descriptions, only the following simple knowledge of the variational method is required.

(1)   A quantity determined by the shape of a function $f(x)$ defined in a given range of the independent variable x is called a functional. In short, a functional is a function of functions.

(2)   A differential equation giving the condition to make a functional stationary (maximum or minimum) is called the Euler's equation for that functional.

(3)   When we solve a differential equation by using the variational method, we first find a functional whose Euler's equation is the given differential equation, and next find the stationary condition of the functional. Thus the differential equation can be solved.

5.5.1   Variational Formulation

We start from three basic equations describing the field in the fibers, Eqs. (5.78), (5.89), and (5.90), which are now given again:

$$\frac{1}{r}\frac{d}{dr}\left(r\frac{dR}{dr}\right) + \left[\omega^2\varepsilon(r)\mu_0 - \beta^2 - \frac{m^2}{r^2}\right]R = 0, \tag{5.178}$$

$$R(a) = R_{\text{clad}}(a), \tag{5.179}$$

$$\left[\frac{dR}{dr}\right]_{r=a} = \left[\frac{dR_{\text{clad}}}{dr}\right]_{r=a}. \tag{5.180}$$

The solution of the differential equation (5.178) under boundary conditions [Eqs. (5.179) and (5.180)] is also given as the stationary condition of a functional:

$$I[R] = \int_0^a \left[\left(\frac{dR}{dr}\right)^2 + \frac{m^2}{r^2}R^2\right]r\,dr - \int_0^a [\omega^2\varepsilon(r)\mu_0 - \beta^2]R^2(r)r\,dr$$
$$- \Phi_\beta R^2(a), \tag{5.181}$$

where Φ_β is defined as

$$\Phi_\beta = \left[\frac{r}{R_{\text{clad}}(r)}\frac{dR_{\text{clad}}}{dr}\right]_{r=a}. \tag{5.182}$$

The proof will be given in the following.

We assume that when $R(r)$ takes an appropriate form, Eq. (5.181) becomes stationary under Eqs. (5.179) and (5.180), and we denote that form of $R(r)$ by $R_0(r)$. Next, using an arbitrary function $\eta(r)$ and small real quantity δ, we write

$$R_\delta(r) = R_0(r) + \delta\eta(r) \tag{5.183}$$

and put this into Eq. (5.181). Then $I[R_\delta]$ is a function of δ. However, according to the definition of R_0, $I[R_\delta]$ is stationary for $\delta = 0$, so that

$$\left[\frac{\partial}{\partial\delta} I[R_\delta]\right]_{\delta=0} = 2\int_0^a r \frac{dR_0}{dr}\frac{d\eta}{dr}\,dr - 2\int_0^a \left[\omega^2\varepsilon(r)\mu_0 - \beta^2 - \frac{m^2}{r^2}\right]rR_0(r)\eta(r)\,dr$$

$$- 2\Phi_\beta R_0(a)\eta(a) = 0. \tag{5.184}$$

Integrating the first term by parts and using Eq. (5.182), we obtain

$$\eta(a)R_0(a)\left\{\frac{a}{R_0(a)}\left[\frac{dR_0}{dr}\right]_{r=a} - \frac{a}{R_{\text{clad}}(a)}\left[\frac{dR_{\text{clad}}}{dr}\right]_{r=a}\right\}$$

$$- \int_0^a \left\{\frac{1}{r}\frac{d}{dr}\left(r\frac{dR_0}{dr}\right) + \left[\omega^2\varepsilon(r)\mu_0 - \beta^2 - \frac{m^2}{r^2}\right]R_0\right\}\eta(r)r\,dr = 0. \tag{5.185}$$

However, since $\eta(r)$ is an arbitrary function of r, the braced portions in the first and second terms must both be zero. If the boundary conditions are satisfied, the first brace automatically becomes zero. The second brace has a form identical to the left-hand side of Eq. (5.178). Thus, it has been proved that the function R_0 making $I[R]$ stationary satisfies both the differential equation and the given boundary conditions.

A comment should be added concerning why the two boundary conditions [Eqs. (5.179) and (5.180)] are reduced to only one as seen in Eq. (5.185). In the present analysis, the amplitude of the field in the cladding region can be given entirely arbitrarily; hence Eq. (5.179) alone gives no actual constraint on the field in the core. The only quantity that has to be made continuous at the core–cladding boundary, therefore, is the ratio of the amplitude and its derivative given in Eq. (5.182).

5.5.2   Dispersion Equation

One of the useful methods for solving a variational problem is the Rayleigh–Ritz method. In this method, the function to be determined [in the present case, $R(r)$] is first expanded by a series of orthogonal functions, and the expansion coefficients are then determined by the stationary condition.

We first note that for the transverse field function in the cladding region $R_{\text{clad}}(r)$, an equation similar to Eq. (5.178) [but $\varepsilon(r)$ being replaced by ε_2] holds. Since $(\omega^2\varepsilon_2\mu_0 - \beta^2) < 0$, the solution in the cladding is given, as described in Section 4.3.1, in the form

$$R_{\text{clad}}(r) = R(a)[K_m(wr/a)/K_m(w)], \tag{5.186}$$

where K_m denotes the second-kind modified Bessel function of the mth order, and w is the normalized transverse wave number in the cladding region [Eq. (4.50)]. From Eqs. (5.182) and (5.186),

$$\Phi_\beta = wK'_m(w)/K_m(w) \tag{5.187}$$

where the prime denotes differentiation with respect to w.

We next introduce a set of orthogonal functions

$$F_{mk}(r) = J_m(\lambda_k r/a)/J_m(\lambda_k) \tag{5.188}$$

to expand $R(r)$. In this expression, λ_k is the kth root of

$$zJ'_m(z) - \Phi_\beta J_m(z) = 0. \tag{5.189}$$

By choosing λ_k in this manner, the boundary conditions [Eqs. (5.179) and (5.180)] are automatically satisfied term by term as seen in the following paragraph. The solution to be obtained is expanded by using $F_{mk}(r)$ as

$$R(r) = \sum_{k=1}^{\infty} C_k F_{mk}(r), \tag{5.190}$$

where the C_k are the constants to be determined. From Bessel-function formulas, it can be shown that $F_{mk}(r)$ satisfies the orthogonality

$$\int_0^a F_{mk}(r)F_{ml}(r)r\,dr = (a^2/2)[1 + (\Phi_\beta^2 - m^2)/\lambda_k^2]\delta_{kl}, \tag{5.191}$$

where λ_{kl} is the Kronecker delta.*

We first investigate the boundary conditions. Differentiating Eq. (5.190) and using Eqs. (5.188) and (5.189), we obtain

$$\frac{a}{R(a)}\left[\frac{dR}{dr}\right]_{r=a} = \frac{\Phi_\beta}{R(a)}\sum_{k=1}^{\infty} C_k = \Phi_\beta, \tag{5.192}$$

so that the boundary condition is automatically satisfied.

Next substitute Eq. (5.190) in (5.181) and perform the integration using Eq. (5.191) and Bessel-function formulas as described in Appendix 5A.5 to obtaining the functional

$$I[C_1, C_2, \ldots] = \tfrac{1}{2}\sum_{k=1}^{\infty}\sum_{l=1}^{\infty} C_k C_l\{(\lambda_k^2 - u^2)[1 + (\Phi_\beta^2 - m^2)/\lambda_k^2]\delta_{kl}$$
$$+ v^2 A_{mkl}\}, \tag{5.193}$$

---

* In this section, l is used to number the orthogonal functions; note that this is different from the radial mode number l used previously.

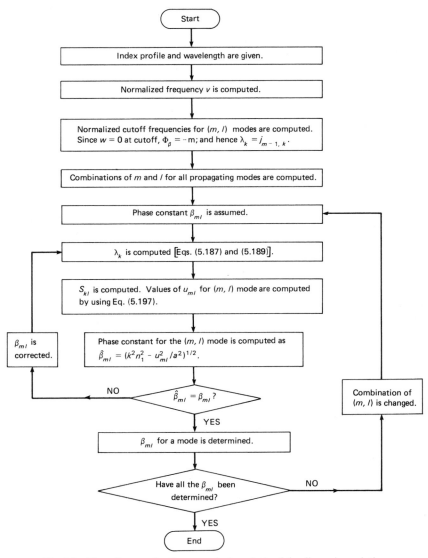

Fig. 5.8   Flow diagram for the numerical analysis of the dispersion relation.

where

$$A_{mkl} = (1/\Delta a^2) \int_0^a f(r) F_{mk}(r) F_{ml}(r) r \, dr, \qquad (5.194)$$

and u, v, and $f(r)$ defined in Eqs. (5.161), (5.162), and (5.101), respectively, are used.

In order that a solution $R(r)$ makes the functional stationary, partial derivatives of $I[C_1, C_2, \ldots]$ with respect to all C_k must be zero; i.e.,

$$\partial I/\partial C_k = \sum_{l=1}^{\infty} C_l S_{kl} = 0 \qquad (k = 1, 2, \ldots), \qquad (5.195)$$

where the coefficients S_{kl} are

$$S_{kl} = (\lambda_k^2 - u^2)[1 + (\Phi_\beta^2 - m^2)/\lambda_k^2] \delta_{kl} + v^2 A_{mkl}. \qquad (5.196)$$

The set of linear homogeneous equations given by Eq. (5.195) has nontrivial solutions for the C_l only if the matrix $\{S_{kl}\}$ is singular, i.e.,

$$\det\{S_{kl}\} = 0. \qquad (5.197)$$

This equation includes the variables u, v, and w. However, w is related to u and v [Eq. (4.77)], whereas u and v determine the propagation constant β [Eq. (4.78)]. Thus, Eq. (5.197) gives the relation between β and v, i.e., the dispersion relation, of an optical fiber having an arbitrary refractive-index profile.

5.5.3   Numerical Analysis of Dispersion Relation

It is not easy to obtain simple solutions of the dispersion relation when the refractive-index profile is arbitrary. When the profile is the α-power type, an approximate but closed-form solution is obtained [11, 12]. For arbitrary profiles, numerical analysis is required.

The flow diagram of the numerical analysis is shown in Fig. 5.8. In this figure $j_{m-1'k}$ denotes the kth root of $J_{m-1}(x) = 0$. Some results from actual computations are shown in Okoshi and Okamoto [10].

5.5.4   Group Delay

A formula expressing group delay has been derived by using the WKB method and is given as Eq. (5.165), but only for α-power profiles. If we use the parameter A_{mkl} used in the Rayleigh–Ritz analysis, we can compute the delay time for an arbitrary profile.

Equation (5.178) gives a proper-value problem for the propagation constant β. We can express β^2, after computations shown in Appendix 5A.6, as

$$\beta^2 = \frac{\int_0^\infty k^2 n^2(r) R^2(r) r\, dr - \int_0^\infty \{(dR/dr)^2 + (m^2/r^2) R^2\} r\, dr}{\int_0^\infty R^2(r) r\, dr}, \quad (5.198)$$

provided $R(r)$ is the solution satisfying Eq. (5.178). An important fact about (5.198) is that, as shown in Appendix 5A.6, β^2 expressed as above is stationary with respect to slight variation of $R(r)$. Hence, Eq. (5.198) is called the variational expression for the propagation constant.

From the foregoing expression for β^2, after computations described in Appendix 5A.7, we obtain

$$\beta \frac{d\beta}{dk} = \frac{\int_0^\infty kn[d(kn)/dk] R^2(r) r\, dr}{\int_0^\infty R^2(r) r\, dr}. \quad (5.199)$$

Further, by substituting Eqs. (5.186) and (5.190) into (5.199) and performing some computations described in Appendix 5A.8, (5.199) is rewritten as

$$\frac{\beta}{kn_1} \frac{d\beta}{dk} = N_1 \left\{ 1 - (1 + \tfrac{1}{4}y) \frac{\int_0^\infty f(r) R^2(r) r\, dr}{\int_0^\infty R^2(r) r\, dr} \right\}$$

$$= N_1 \left\{ 1 - (2 + \tfrac{1}{2}y)\Delta \frac{\sum_{k=1}^\infty \sum_{l=1}^\infty C_k C_l [A_{mkl} + 1/\xi_m - 1]}{\sum_{k=1}^\infty \sum_{l=1}^\infty C_k C_l [\{1 + (\Phi_\beta^2 - m^2)/\lambda_k^2\} \delta_{kl} + 1/\xi_m - 1]} \right\}, \quad (5.200)$$

where ξ_m is a quantity defined by Eq. (4.103),

$$\xi_m = K_m^2(w)/K_{m-1}(w) K_{m+1}(w). \quad (5.201)$$

Therefore, if we define and use

$$\Theta = \frac{\sum_{k=1}^\infty \sum_{l=1}^\infty C_k C_l [A_{mkl} + 1/\xi_m - 1]}{\sum_{k=1}^\infty \sum_{l=1}^\infty C_k C_l [\{1 + (\Phi_\beta^2 - m^2)/\lambda_k^2\} \delta_{kl} + 1/\xi_m - 1]}, \quad (5.202)$$

the delay time for unit length of fiber t is expressed in a simple form,

$$t = \frac{1}{c} \frac{d\beta}{dk} = \frac{N_1}{c} \frac{[1 - \Delta(2 + \tfrac{1}{2}y)\Theta]}{(1 - 2\Delta u^2/v^2)^{1/2}}. \quad (5.203)$$

In actual computations, we first determine u from solutions of Eq. (5.197), and then C_{kl} as solutions of Eq. (5.195). Substituting these values into Eq. (5.202), we can compute Θ. The delay time t is then computed without difficulty by using Eq. (5.203).

The Rayleigh–Ritz method will be used in Section 5.7.2 in the preliminary investigation of the optimum refractive-index profile for multimode fibers, and also in Section 7.3 in the computer-aided "synthesis" of that optimum index profile.

5.6 Analysis of Fibers Having α-Power Index Profiles by Power-Series Expansion Method

5.6.1 Introduction

For analysis of the propagation characteristics of optical fibers having really arbitrary refractive-index profiles, various computer analyses are available as described in the preceding and following sections. However, all of these analyses lead finally to matrix-type dispersion formulas which can never be solved without relying fully on a computer.

For somewhat restricted cases in which the index profile is expressed as the α power of the radial coordinate, three simpler approaches are possible. The first is the WKB analysis described in Section 5.4. However, in the WKB analysis, the presence of many propagating modes is assumed, and the effect of those modes close to cutoff is neglected [2]. Besides, the effect of the index valley at the core–cladding boundary, which plays an important role in reducing multimode dispersion, cannot be treated by ordinary WKB analysis such as described in Section 5.4.

The second example is the closed-form approximate dispersion formula for α-power nonuniform-core fibers [11, 12], which has been derived from the Rayleigh–Ritz formulation described in the preceding section.* This simple approximate formula is also applicable to the close-to-cutoff modes, and can be used to evaluate the effect of the index valley for the reduction of multimode dispersion. However, the derivation of this formula is omitted here because more exact and relatively simple solutions can be given by the third, following method.

The third approach is the power-series expansion method [13]. This method should essentially be applicable to arbitrary profiles, but is most effectively used for α-power profiles. In the following the dispersion and delay-time equations are derived by this method. Propagation characteristics of α-power nonuniform-core fibers are also investigated using these equations. It will be shown that a fiber with a certain refractive-index profile exhibits minimal multimode dispersion.

---

* This dispersion formula was first derived incorrectly as an exact formula within the scalar-wave approximation, because an implied assumption is overlooked [11]. The correct derivation of the same formula, but as an approximate formula, was found later [12].

5.6.2   α-Power Profile with Step or Valley at Core–Cladding Boundary

We consider the α-power refractive-index profile

$$n(r) = \begin{cases} n_1[1 - 2\rho\,\Delta(r/a)^{\alpha}]^{1/2} & (0 \leqq r \leqq a), & (5.204a) \\ n_2 = [n_1 - 2\Delta]^{1/2}. & (a < r), & (5.204b) \end{cases}$$

where a denotes the core radius, n_1 and n_2 the refractive indices of the axis and in the cladding, respectively, Δ the relative refractive-index difference between the core axis and cladding [as in Eq. (5.155)], ρ a parameter representing the refractive-index step or valley at the core–cladding boundary. A smooth continuation at the core–cladding boundary, the presence of a step, and that of a valley are expressed by $\rho = 1$, $\rho < 1$, and $\rho > 1$, respectively (see Fig. 5.9). Note that Eq. (5.204) is a little different from the α-power profile investigated by the WKB analysis [Eq. (5.155)] where $\rho = 1$ throughout.

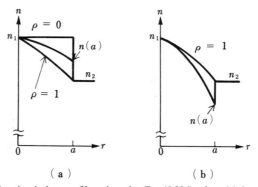

Fig. 5.9   The refractive-index profiles given by Eq. (5.204) when (a) $0 \leq \rho \leq 1$, (b) $1 \leq \rho$.

5.6.3   Derivation of Dispersion Relation

We start from Eq. (5.178). Assuming the preceding index profile and using the definitions of u [Eq. (5.161)], v [Eq. (5.162)], and a normalized variable

$$x = r/a, \tag{5.205}$$

we can rewrite Eq. (5.178) as

$$\frac{1}{x}\frac{d}{dx}\left(x\frac{dR}{dx}\right) + \left(u^2 - \rho v^2 x^{\alpha} - \frac{m^2}{x^2}\right)R = 0. \tag{5.206}$$

If we assume here that in the core ($x \leq 1$)

$$R = R_0 x^\lambda \sum_{n=0}^{\infty} a_n x^n, \tag{5.207}$$

we obtain

$$\sum_{n=0}^{\infty} (n + \lambda)^2 a_n x^{n-2} + u^2 \sum_{n=0}^{\infty} a_n x^n - \rho v^2 \sum_{n=0}^{\infty} a_n x^{n+\alpha} - m^2 \sum_{n=0}^{\infty} a_n x^{n-2} = 0.$$

$$\tag{5.208}$$

Comparison of coefficients of each power leads to

(1)   from x^{-2} terms: $(\lambda^2 - m^2)a_0 = 0$;   (5.209)
(2)   from x^{-1} terms: $\{(\lambda + 1)^2 - m^2\}a_1 = 0$;   (5.210)
(3)   from x^{n-2} terms (where $n \geq 2$):

$$\{(n + \lambda)^2 - m^2\}a_n + u^2 a_{n-2} = 0 \quad \text{(for } 2 \leq n < \alpha + 2), \tag{5.211}$$

$$\{(n + \lambda)^2 - m^2\}a_n + u^2 a_{n-2} - \rho v^2 a_{n-\alpha-2} = 0 \quad \text{(for } \alpha + 2 \leq n). \tag{5.212}$$

From Eq. (5.209), in order that $a_0 \neq 0$, $\lambda = \pm m$. However, since $R(x)$ should not diverge at $x = 0$, the negative sign is insignificant. Therefore,

$$\lambda = m. \tag{5.213}$$

We can assume, since a_0 is finite but arbitrary,

$$a_0 = 1, \tag{5.214}$$

and from Eq. (5.210),

$$a_1 = 0. \tag{5.215}$$

Further, from Eqs. (5.211) and (5.212),

$$a_n = \begin{cases} -[n(n + 2m)]^{-1} u^2 a_{n-2} & (2 \leq n < \alpha + 2), \tag{5.216a} \\ -[n(n + 2m)]^{-1}(u^2 a_{n-2} - \rho v^2 a_{n-\alpha-2}) & (\alpha + 2 \leq n), \tag{5.216b} \end{cases}$$

and $a_n = 0$ for negative n. The field function $R(x)$ can be determined by using these recurrence formulas.

The dispersion relation (proper equation) is derived from the boundary condition. From the continuity of R and dR/dx at $x = 1$,

$$\frac{1}{R(1)} \left[\frac{dR}{dx} \right]_{x=1} = \frac{1}{K_m(w)} \left[\frac{d}{dx} K_m(wx) \right]_{x=1}, \tag{5.217}$$

the proper equation is

$$\frac{\sum_{n=0}^{\infty} n a_n}{\sum_{n=0}^{\infty} a_n} + 2m + \frac{w K_{m-1}(w)}{k_m(w)} = 0. \tag{5.218}$$

This is the dispersion equation giving, together with Eqs. (5.214)–(5.216) and the u–v–w relation, the propagation characteristics of α-power non-uniform-core fibers.

5.6.4   Cutoff Conditions

At cutoff frequencies for each mode,

$$u = v, \qquad (5.219)$$

$$w = 0. \qquad (5.220)$$

Hence, Eq. (5.218) becomes

$$\left(\sum_{n=0}^{\infty} nb_n \Bigg/ \sum_{n=0}^{\infty} b_n \right) + 2m = 0, \qquad (5.221)$$

where

$$b_0 = 1, \qquad (5.222)$$

$$b_1 = 0, \qquad (5.223)$$

$$b_n = \begin{cases} -\left[n(n+2m)\right]^{-1} v^2 b_{n-2}, & (5.224a) \\ -\left[n(n+2m)\right]^{-1} v^2 (b_{n-2} - \rho b_{n-\alpha-2}). & (5.224b) \end{cases}$$

The normalized cutoff frequencies v_c can be determined by solving Eqs. (5.221)–(5.224); the lth smallest solution of v gives v_c for the LP$_{ml}$ mode. The preceding cutoff-frequency formula was derived first by Gambling *et al.* [14], but only for the case $m = 0$.

5.6.5   Group Delay

When the dispersion relation (k–β relation) is obtained in the form of $f(k, \beta) = 0$, the group delay is

$$t = -\frac{1}{c} \frac{\partial f / \partial k}{\partial f / \partial \beta}. \qquad (5.225)$$

From Eqs. (5.218) and (5.225), this can be rewritten as

$$t = -c^{-1} \left(\frac{\partial}{\partial k} \left[\frac{\sum na_n}{\sum a_n} + w \frac{K_{m-1}(w)}{K_m(w)} \right] \Bigg/ \frac{\partial}{\partial \beta} \left[\frac{\sum na_n}{\sum a_n} + w \frac{K_{m-1}(w)}{K_m(w)} \right] \right). \qquad (5.226)$$

5.6.6   Numerical Solutions

To obtain the dispersion characteristics, Eqs. (5.218), (5.214)–(5.216), the u–v–w relation [Eq. (4.77)], and the u–β relation [Eq. (5.161)] are used to compute the β–v relation for each mode. The lth smallest solution of v for a given β gives v for the LP$_{ml}$ mode. The group-delay characteristics are then obtained from (5.226).

Figures 5.11a–5.11f show the dispersion and group-delay characteristics (curve-plotter output) thus obtained for the six α-power profiles shown in Fig. 5.10 [13]. The upper and lower figures for each case in Fig. 5.11 show

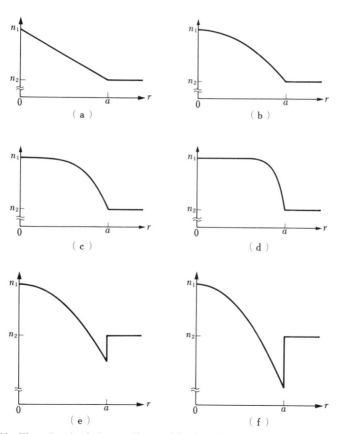

Fig. 5.10   The refractive-index profiles used in the calculations of dispersion and group-delay characteristics shown in Fig. 5.11. (a) $\alpha = 1$, $\rho = 1.0$; (b) $\alpha = 2$, $\rho = 1.0$; (c) $\alpha = 4$, $\rho = 1.0$; (d) $\alpha = 10$, $\rho = 1.0$; (e) $\alpha = 2$, $\rho = 1.5$; and (f) $\alpha = 2$, $\rho = 2.0$.

the group-delay and dispersion characteristics, respectively. The ordinate of the upper figures gives the relative group-delay difference,

$$T = (ct/N_1) - 1, \tag{5.227}$$

where c denotes light velocity, t the delay per unit length of fiber, and N_1 is the group index of the material used at the core center. This ordinate is essentially identical to that in Fig. 4.9. The ordinate of the lower figures gives the normalized propagation constant defined in Eq. (4.82):

$$x \equiv (k^2 n_1^2 - \beta^2)/(k^2 n_1^2 - k^2 n_2^2) \simeq (kn_1 - \beta)/(kn_1 - kn_2), \tag{5.228}$$

where k denotes the vacuum wave number, β the propagation constant, and n_1 and n_2 the refractive indices of the fiber axis and in the cladding, respectively. This ordinate is identical to that in Fig. 4.3. [Note that x is used in Eqs. (5.205) and (5.228) to denote different quantities.] The abscissa v is the conventional normalized frequency defined in Eq. (4.76).

In all the figures, Δ [see Eq. (5.204)] is assumed to be 0.01, and the index profile is characterized by two parameters α and ρ. The error in these curves has been investigated by comparing them with the results of the finite-element method analyses (Section 5.8), and the relative error in the horizontal direction is estimated to be below 1% for both T and x [13].

Exact values of cutoff frequencies are often useful because the accuracy of a newly developed analysis technique can be estimated by comparing the obtained cutoff frequencies with the exact ones. For such a purpose, the exact cutoff frequencies computed by using Eqs. (5.221)–(5.224) for $\alpha = 1, 2, 4$, and 10, $\rho = 1$, $m = 0$–9, and $l = 1$–10 are tabulated in Table 5A.1 (Appendix 5A.9). To compute these values, the "quadruple precision" (33-digit) computation was performed to assure accuracy; comparison of these values with the results of finite-element method analyses suggests that the relative error $\Delta v_c/v_c$ is less than 5×10^{-5} except for $LP_{8,10}$ and $LP_{9,10}$ modes for $\alpha = 1$, where the error exceeds this limit a little [13].

5.7 Suggestions on the Optimum Index Profile for Multimode Fibers

5.7.1 Optimum Values of α and ρ

Figure 5.11f shows that when $\alpha = 2$ and $\rho = 2$, the delay times for most modes are concentrated around $t = N_1/c$ (i.e., $T = 0$). This fact suggests that a very low multimode dispersion can be realized in this condition. In Sections 5.4.2 and 5.4.3, the optimum value of α has been discussed on the basis of the WKB approximation, and it was shown [Eq. (5.166)] that when

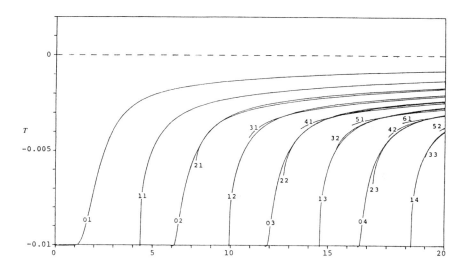

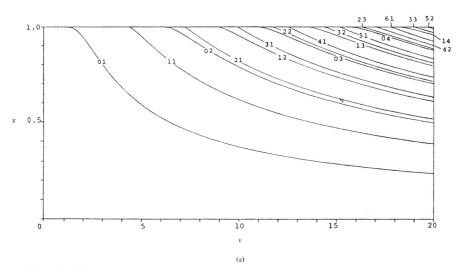

(a)

Fig. 5.11   Group-delay characteristics (upper figures) and dispersion characteristics (lower figures) of various α-power nonuniform-core fibers computed by power-series expansion method (after Okoshi and Oyamada [13]). (a) $\alpha = 1$, $\rho = 1.0$, (b) $\alpha = 2$, $\rho = 1.0$, (c) $\alpha = 4$, $\rho = 1.0$, (d) $\alpha = 10$, $\rho = 1.0$, (e) $\alpha = 2$, $\rho = 1.5$, and (f) $\alpha = 2$, $\rho = 2.0$. The numerals in the figure denote the LP-mode numbers as in Figs. 4.3 and 4.9.

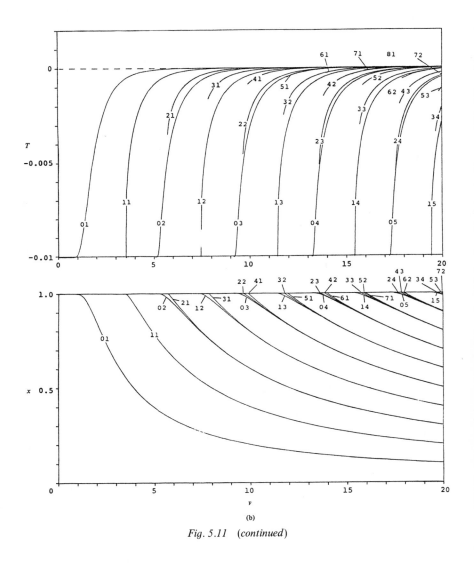

Fig. 5.11   (continued)

$\alpha = 2 + y$, the dispersion could much be reduced compared with that of uniform-core fibers. Comparison of Figs. 5.11b ($\alpha = 2$, $\rho = 1$) and 5.11f ($\alpha = 2$, $\rho = 2$) suggests that an appreciable, further reduction can be expected by providing a valley having an appropriate depth at the core–cladding boundary. (Note that in these cases we assume $y = 0$.)

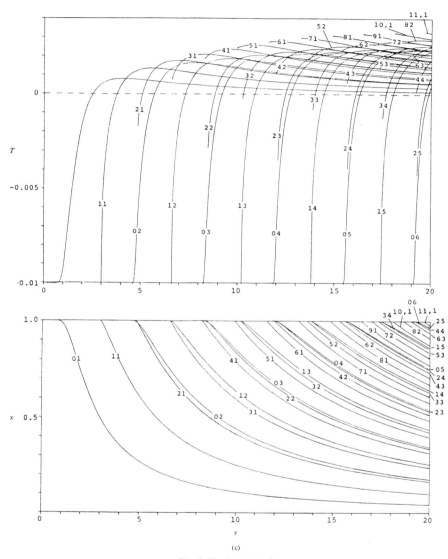

(c)

Fig. 5.11   (continued)

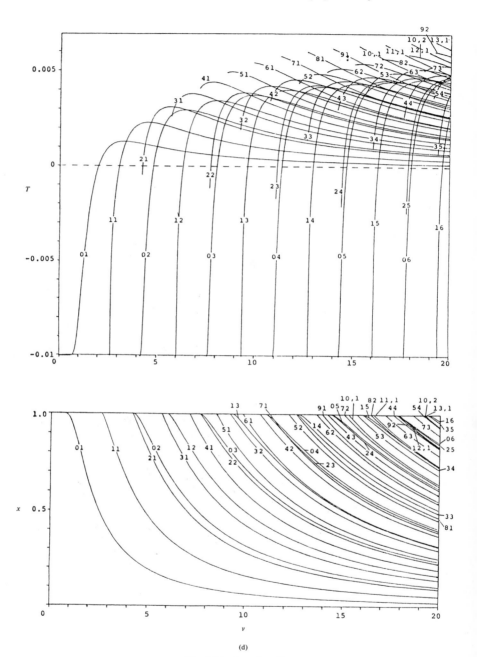

Fig. 5.11   (continued)

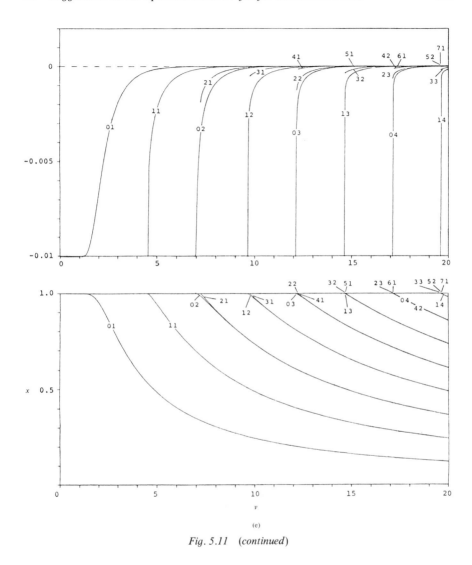

Fig. 5.11   (continued)

5.7.2   Effect of Index Valley at Core–Cladding Boundary

To investigate in more detail the effect of the index valley on the reduction of multimode dispersion, the magnitude of the dispersion has been computed by using the Rayleigh–Ritz method as a function of the depth and shape of the valley [15]. It is assumed throughout that $y = 0$ and $\alpha = \alpha_{opt} = 2$.

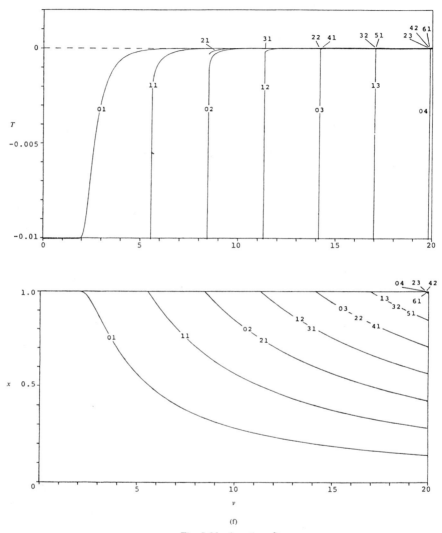

Fig. 5.11   (*continued*)

We first define the variance (mean square spread) of the intermodel delay difference at a normalized frequency v as

$$\sigma(v) = N^{-1} \sum_{i=1}^{N} \left[t_i(v) - \langle t(v) \rangle \right]^2, \tag{5.229}$$

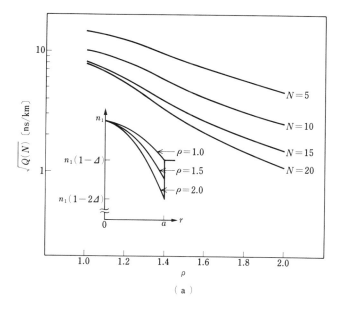

(a)

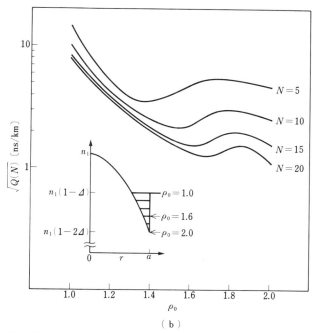

(b)

Fig. 5.12   Variation of rms multimode dispersion for various depths and shapes of the index valley (after Okamoto and Okoshi [15]) as functions of (a) ρ, and (b) ρ_0 (the level to which the valley is filled).

where

$$\langle t(v) \rangle = N^{-1} \sum_{i=1}^{N} t_i(v). \tag{5.230}$$

In these equations, N denotes the number of propagating LP modes at frequency v, $t_i(v)$ is the group delay of the ith mode, and $\langle t(v) \rangle$ denotes, as seen in Eq. (5.230), the average $t_i(v)$ with respect to i.

We consider a range of the normalized frequency v within which N LP-modes can propagate. If the normalized cutoff frequency of the Nth mode is denoted by v_{cN}, the value of $\sigma(v)$ will be remarkably different at v a little greater than v_{cN} and v a little lower than $V_{c(N+1)}$ (see Fig. 5.11). Hence, as a reasonable measure of the multimode dispersion, we introduce a new quantity obtained by averaging $\sigma(v)$:

$$Q(N) = \frac{\int_{v_{cN}+\varepsilon}^{v_{c(N+1)}-\varepsilon} \sigma(v)\,dv}{(v_{c(N+1)} - \varepsilon) - (v_{cN} + \varepsilon)}, \tag{5.231}$$

where ε is an appropriately small quantity introduced to exclude the drastic variation of $\sigma(v)$ near cutoff. It will be found in Chapter 7 that relation between $Q(N)$ and the maximum intermodal delay difference T is given approximately as $\sqrt{Q(N)} = \frac{1}{4} T$ (Fig. 7.8).

The variation of $\sqrt{Q(N)}$ for various depths and shapes of the index valley is shown in Fig. 5.12, the number of the propagating LP modes N being the parameter. Figure 5.12a, which shows $\sqrt{Q(N)}$ as a function of ρ, suggests that $\sqrt{Q(N)}$ is reduced by almost one order of magnitude by increasing ρ from 1.0 to 2.0. Figure 5.12b, which shows $\sqrt{Q(N)}$ as a function of the level ρ_0 to which the valley is filled, suggests that $\sqrt{Q(N)}$ varies in a complicated manner as ρ_0 increases, but in many cases exhibits a minimum at a certain value of ρ_0. These observations substantiate the validity of the "computer-synthesized" optimum refractive-index profile to be shown in Chapter 7.

5.8   Analysis of Nonuniform-Core Fibers by Finite Element Method

5.8.1   Introduction

In the analysis of transmission characteristics of optical fibers, the scalar wave approximation is often made because it simplifies the analysis remarkably; the resulting error is tolerable in most cases. To obtain the error estimate, however, the result of the analysis must be compared with that of rigorous, vectorial wave analysis.

So far three methods of the scalar wave analysis of nonuniform-core fibres have been presented: (1) WKB method; (2) Rayleigh–Ritz method; and last, (3) power-series expansion method. Two additional methods of the vectorial wave analysis of nonuniform-core fibres will be described: (1) finite element method (this section); and (2) staircase-approximation method (Section 5.9). Later in this section, results of the vectorial wave and scalar wave analyses are compared for a wide variety of refractive-index profiles. It will be found that the errors caused by the scalar wave approximation in the propagation constant and group delay are about 0.1 and 1%, respectively, when the relative-index difference $\Delta = 0.01$ [16].

5.8.2 Variational Formulation

The starting equations are the eight equations derived in Section 5.2.4 [Eqs. (5.29)–(5.36)], which are now cited for convenience:

$$\left(\frac{\chi - f}{1 - f}\right)\frac{1}{r}\frac{d}{dr}\left[\left(\frac{1 - f}{\chi - f}\right)r\frac{d\Phi}{dr}\right] + \left[\omega^2\varepsilon_1\mu_0(\chi - f) - \frac{n^2}{r^2}\right]\Phi$$

$$+ \frac{n}{r}\left(\frac{\chi - f}{1 - f}\right)\Psi\frac{d}{dr}\left(\frac{1 - f}{\chi - f}\right) = 0, \qquad (5.232)$$

$$(\chi - f)\frac{1}{r}\frac{d}{dr}\left[\frac{1}{\chi - f}r\frac{d\Psi}{dr}\right] + \left[\omega^2\varepsilon_1\mu_0(\chi - f) - \frac{n^2}{r^2}\right]\Psi$$

$$+ \frac{n}{r}(\chi - f)\Phi\frac{d}{dr}\left(\frac{1}{\chi - f}\right) = 0, \qquad (5.233)$$

$$E_z = (\omega^2\varepsilon_1\mu_0/\beta)\Phi(r)\cos(n\theta + \varphi_n), \qquad (5.234)$$

$$H_z = \omega\varepsilon_1\Psi(r)\sin(n\theta + \varphi_n), \qquad (5.235)$$

$$E_r = -j\frac{1}{\chi - f}\left[\frac{d\Phi}{dr} + \frac{n}{r}\Psi\right]\cos(n\theta + \varphi_n), \qquad (5.236)$$

$$E_\theta = j\frac{1}{\chi - f}\left[\frac{d\Psi}{dr} + \frac{n}{r}\Phi\right]\sin(n\theta + \varphi_n), \qquad (5.237)$$

$$H_r = -j\frac{\beta}{\omega\mu_0}\frac{1}{\chi - f}\left[\frac{d\Psi}{dr} + \left(\frac{1 - f}{1 - \chi}\right)\frac{n}{r}\Phi\right]\sin(n\theta + \varphi_n), \qquad (5.238)$$

$$H_\theta = -j\frac{\beta}{\omega\mu_0}\frac{1}{\chi - f}\left[\left(\frac{1 - f}{1 - \chi}\right)\frac{d\Phi}{dr} + \frac{n}{r}\Psi\right]\cos(n\theta + \varphi_n), \qquad (5.239)$$

where n denotes an integer, $\varphi_n = 0$ or $\pi/2$, and

$$f(r) = 1 - \varepsilon(r)/\varepsilon_1, \tag{5.240}$$

$$\chi = 1 - \beta^2/\omega^2\varepsilon_1\mu_0. \tag{5.241}$$

We manipulate the problem into a variational form. As described in Appendix 5A.10, the solutions Φ and Ψ which satisfy the foregoing vectorial wave equations (Eqs. (5.232) and (5.233)] and the boundary conditions (continuity of E_z, H_z, E_r, E_θ, H_r, and H_θ at the core–cladding boundary $r = a$) may also be obtained as that solution of the variational problem to make the following functional sationary:

$$
\begin{aligned}
I[\Phi, \Psi] = {} & \frac{1}{1-\chi} \int_0^a \frac{1-f}{\chi-f}\left[\left(\frac{d\Phi}{dr}\right)^2 + \frac{n^2}{r^2}\Phi^2\right] r\,dr - \frac{k^2 n_1^2}{1-\chi}\int_0^a (1-f)\Phi^2 r\,dr \\
& + \int_0^a \frac{1}{\chi-f}\left[\left(\frac{d\Psi}{dr}\right)^2 + \frac{n^2}{r^2}\Psi^2\right]r\,dr - k^2 n_1^2 \int_0^a \Psi^2 r\,dr \\
& + \int_0^a \frac{2n}{\chi-f}\frac{d}{dr}(\Phi\Psi)\,dr - \frac{1}{\chi-2\Delta}\left[\Omega_\beta\frac{1-2\Delta}{1-\chi}\Phi^2(a)\right. \\
& \left. + 2n\Phi(a)\Psi(a) + \Omega_\beta\Psi^2(a)\right]
\end{aligned}
\tag{5.242}
$$

where

$$w = (\beta^2 - k^2 n_2^2)^{1/2} a, \tag{5.243}$$

$$\Omega_\beta = w K_n'(w)/K_n(w), \tag{5.244}$$

and K_n denotes the nth-order modified Bessel function of the second kind.

5.8.3 Solution of the Variational Problem by Finite Element Method

The Rayleigh–Ritz method has been used to solve variational problems in the preceding sections. However, in the present vectorial wave analysis this method cannot be used for the following reason. From Eqs. (5.240) and (5.241),

$$\chi - f = [n^2(r) - \beta^2/k^2]/n_1^2. \tag{5.245}$$

Since β^2/k^2 varies in the range $n_2^2 < \beta^2/k^2 < n_1^2$ for propagating modes, $(\chi - f)$ becomes zero at one point $(r = r_p)$ as shown in Fig. 5.13. The Rayleigh–Ritz method cannot be easily applied to the present problem because the term $(\chi - f)$ in the denominators in Eq. (5.242) becomes zero at $r = r_p$, where the integrand diverges.

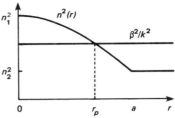

Fig. 5.13 Relation between the refractive-index profile $n(r)$ and the normalized propagation constant β/k.

This difficulty can be avoided by using the finite element method [16]. We first divide the region between $r = 0$ and $r = a$ into N elements (see Fig. 5.14), and express the values of $\Phi(r)$ and $\Psi(r)$ at $r = r_l$ as

$$\Phi_l = \Phi(r_l), \qquad \Psi_l = \Psi(r_l) \qquad (l = 0, 1, 2, \ldots, N). \tag{5.246}$$

Here the division is made so that r_p coincides with an r_l. In each of these elements, the functions $\phi(r)$ and $\psi(r)$ are expressed as

$$\Phi(r) = \Phi_{l-1}F_{l-1}(r) + \Phi_l F_l(r) \tag{5.247}$$
$$\Psi(r) = \Psi_{l-1}F_{l-1}(r) + \Psi_l F_l(r) \tag{5.248}$$
$$(r_{l-1} \leq r \leq r_l)$$

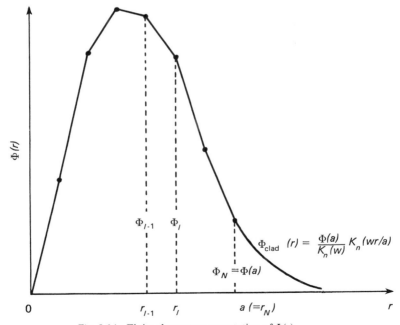

Fig. 5.14 Finite element representation of $\Phi(r)$.

where $F_{l-1}(r)$ and $F_l(r)$ are continuous functions of r satisfying the conditions

$$F_{l-1}(r_{l-1}) = 1, \qquad F_{l-1}(r_l) = 0,$$
$$F_l(r_{l-1}) = 0, \qquad F_l(r_l) = 1. \tag{5.249}$$

In those elements which do not include the singular point ($r = r_p$), the function $F_l(r)$ is approximated, as is done in most finite element analyses [17], by a linear function as shown in Fig. 5.14. In those elements including r_p, $\Phi(r)$ and $\Psi(r)$ are approximated, to avoid the divergence of the integral, as

$$\Phi(r) = (Ar^{-n} + Br^n) - (Cr^{-n} + Dr^n)(\chi - f)^2, \tag{5.250}$$

$$\Psi(r) = (Ar^{-n} - Br^n) - (Cr^{-n} - Dr^n)(\chi - f)^2, \tag{5.251}$$

where n is the azimuthal mode number, and constants A, B, C, D are determined so that $\Phi(r)$ and $\Psi(r)$ coincide with Φ_p and Ψ_p at $r = r_p$ and with Φ_{p-1} and Ψ_{p-1} at $r = r_{p-1}$ (or Φ_{p+1} and Ψ_{p+1} at $r = r_{p+1}$), respectively. (Here the subscript p is used as the numbering of the sampling point coinciding with r_p.)

To make the functional I stationary with respect to all the parameters Φ_l and Ψ_l, the following conditions must hold for all l:

$$\partial I/\partial\Phi_l = 0, \qquad \partial I/\partial\Psi_l = 0. \tag{5.252}$$

Substituting Eqs. (5.247)–(5.251) into (5.242) and using the stationary conditions [Eq. (5.252)], we obtain a matrix equation of the form

$$\begin{bmatrix} \mathbf{S} & \mathbf{T} \\ \mathbf{T} & \mathbf{P} \end{bmatrix} \begin{bmatrix} \mathbf{\Phi} \\ \mathbf{\Psi} \end{bmatrix} = 0, \tag{5.253}$$

where $\mathbf{\Phi} = [\Phi_0, \Phi_1, \ldots, \Phi_N]^T$ (T denoting transposition), $\mathbf{\Psi} = [\Psi_0, \Psi_1, \ldots, \Psi_N]^T$, and $\mathbf{S}, \mathbf{T}, \mathbf{P}$ are $(N+1) \times (N+1)$ matrices whose elements are (for space limitations only some typical elements are shown as examples)

$$S_{l,l-1} = -\frac{k^2 n_1^2}{1-\chi} \int_{r_{l-1}}^{r_l} (1-f)F_{l-1}F_l r\, dr$$

$$+ \frac{1}{1-\chi} \int_{r_{l-1}}^{r_l} \frac{1-f}{\chi-f}\left[\frac{dF_{l-1}}{dr}\frac{dF_l}{dr} + \frac{n^2}{r^2}F_{l-1}F_l\right] r\, dr, \tag{5.254}$$

$$T_{l,l-1} = n \int_{r_{l-1}}^{r_l} \frac{1}{\chi-f}\frac{d}{dr}(F_{l-1}F_l)\, dr, \tag{5.255}$$

$$P_{l,l-1} = -k^2 n_1^2 \int_{r_{l-1}}^{r_l} F_{l-1}F_l r\, dr$$

$$+ \int_{r_{l-1}}^{r_l} \frac{1}{\chi-f}\left[\frac{dF_{l-1}}{dr}\frac{dF_l}{dr} + \frac{n^2}{r^2}F_{l-1}F_l\right] r\, dr. \tag{5.256}$$

In order for a nontrivial solution of Eq. (5.253) to exist,

$$\begin{vmatrix} \mathbf{S} & \mathbf{\Psi} \\ \mathbf{T} & \mathbf{P} \end{vmatrix} = 0. \tag{5.257}$$

This equation is the rigorous proper equation (dispersion equation) which determines the propagation constants of an inhomogeneous optical fiber.

5.8.4   Scalar Wave Approximation

Since for practical fibers, χ and $|f(r)|$ are much smaller than unity, we may approximate in Eqs. (5.232) and (5.233)

$$1 - f(r) \simeq 1, \qquad 1 - \chi \simeq 1. \tag{5.258}$$

Under such approximations, as shown by Yamada and Inaba [18], adding Eqs. (5.232) and (5.233) or subtracting Eq. (5.232) from (5.233), we can obtain the scalar wave equations (see Appendix 5A.11). Hence, Eq. (5.258) gives the scalar wave approximations. If we put these approximations into Eq. (5.242), the functional will be simplified as

$$
\begin{aligned}
I[\Phi, \Psi] = {}& \int_0^a \frac{1}{\chi - f} \left[\left(\frac{d\Phi}{dr} \right)^2 + \frac{n^2}{r^2} \Phi^2 \right] r\, dr - k^2 n_1^2 \int_0^a \Phi^2 r\, dr \\
& + \int_0^a \frac{1}{\chi - f} \left[\left(\frac{d\Psi}{dr} \right)^2 + \frac{n^2}{r^2} \Psi^2 \right] r\, dr - k^2 n_1^2 \int_0^a \Psi^2 r\, dr \\
& + \int_0^a \frac{2n}{\chi - f} \frac{d}{dr} (\Phi\Psi) r\, dr \\
& - \frac{1}{\chi - 2\Delta} \left[\Omega_\beta \Phi^2(a) + 2n\Phi(a)\Psi(a) + \Omega_\beta \Psi^2(a) \right].
\end{aligned} \tag{5.259}
$$

By solving this variational problem by the aforementioned finite element method, we obtain the proper equation corresponding to Eq. (5.257) but having the simpler form

$$\begin{vmatrix} \mathbf{P} & \mathbf{T} \\ \mathbf{T} & \mathbf{P} \end{vmatrix} = 0. \tag{5.260}$$

This is the dispersion equation obtained with the scalar wave approximation.

5.8.5   Numerical Results

The accuracy of the finite-element-method analysis itself is investigated first. The normalized cutoff frequency v of the TE_{01} mode in a uniform-core fiber is computed by the finite element method and compared with the

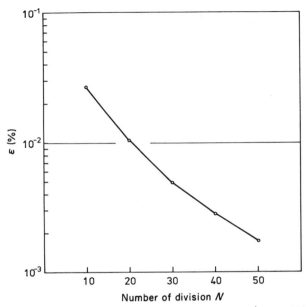

Fig. 5.15   Accuracy of the finite element method: $\varepsilon = 1v_c - j_{0,1}|/j_{0,1} \times 100\%$. Parameter v_c is the normalized cutoff frequency of a uniform-core fiber for the TE_{01} mode obtained by the FEM, whereas $j_{0,1}(=2.4048256\ldots)$ is the exact solution (after Okamoto and Okoshi [16]).

exact value: $j_{0,1} = 2.4048256$. The accuracy thus estimated is shown in Fig. 5.15 as a function of the number of divisions in the finite-element-method analysis.

Next, the error caused by the scalar wave approximation is investigated by comparing the results of analyses using Eqs. (5.257) and (5.260). For this purpose we consider again the α-power index profiles in the core region,

$$n^2(r) = n_1^2[1 - 2\rho\Delta(r/a)^\alpha] \qquad (0 \le r \le a), \qquad (5.261)$$

where Δ denotes the relative index difference between core and cladding and ρ is a parameter representing the presence of a step or valley at core–cladding boundary (see Section 5.6.2).

Figures 5.16a and 5.16b show the error in the normalized frequency v caused by the scalar wave approximation as functions of α and ρ, respectively.* The term "error in the normalized frequency" deserves comments.

---

* One might question why the error in v is shown instead of that in β (or x). It is simply because the error appears in such a form that the vx curve is translated in the horizontal direction (parallel to the v axis) without remarkably changing its overall shape.

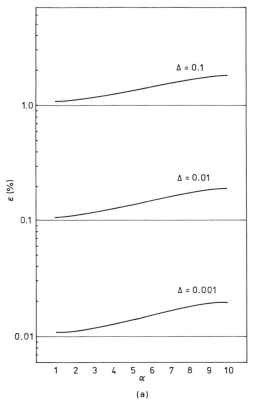

(a)

Fig. 5.16 Error in the normalized frequency caused by the scalar wave approximation: as functions of (a) $\alpha[\rho = 1.0, \varepsilon = |v - v_0|/v_0 \times 100]$ and (b) $\rho[\alpha = 2, \varepsilon = |v - v_0|/v_0 \times 100]$ (after Okamoto and Okoshi [16]).

In the present estimation of the "error," for the sake of convenience, the relative errors [the relative difference between solutions of Eqs. (5.257) and (5.260)] in the normalized frequency v which give three specific propagation constants β for TM_{01} mode are first computed. (The TE_{01} mode is not employed because the scalar wave analysis is exact for this mode when $\alpha = \infty$, i.e., in uniform-core fibers. Refer to Section 4.3.3.) The specific propagation constants are chosen as $x = 0.25$, 0.50, and 0.75, where x is the normalized parameter representing the propagation constant as defined in Eq. (5.228) which is

$$x \equiv (k^2 n_1^2 - \beta^2)/(k^2 n_1^2 - k^2 n_2^2). \tag{5.262}$$

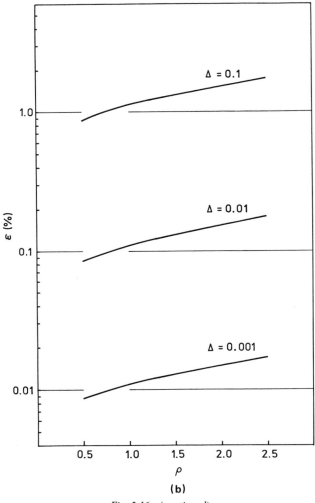

Fig. 5.16   (continued)

The computed errors are then averaged for the preceding three x values to obtain ε shown in the ordinate of Figs. 5.16a and 5.16b. It is found that when $\Delta = 0.01$, which is the typical index difference in multimode fibers, the error caused by the scalar wave approximation is typically 0.1%.

Next, the error caused by the scalar wave approximation in the group delay is estimated. As a preliminary step, we first compute the exact group

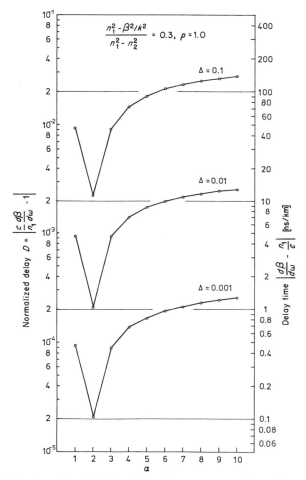

Fig. 5.17   Variation of the delay time for various powers α and $\rho = 1.0$ at $x = 0.3$ (exact solution) (after Okamoto and Okoshi [16]).

delay using the vectorial wave analysis. Figure 5.17 shows the exact normalized delay difference defined by

$$D = \left| \frac{c}{n_1} \frac{d\beta}{d\omega} - 1 \right| \qquad (c = \text{light velocity}) \qquad (5.263)$$

as a function of the power α for the TM_{01} mode at a frequency where $x = 0.3$. Note that D is a parameter proportional to the delay difference between

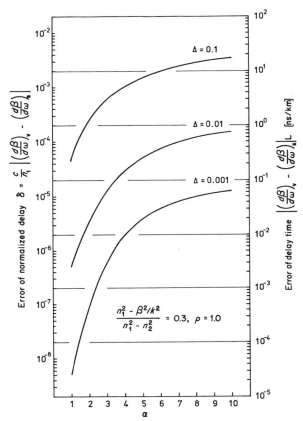

Fig. 5.18   Error in the delay time caused by the scalar wave approximation for $\rho = 1.0$ at $x = 0.3$. Derivatives $(d\beta/d\omega)_v$ and $(d\beta/d\omega)_s$ denote the delay time per unit length obtained by the vectorial wave and scalar wave analyses, respectively (after Okamoto and Okoshi [16]).

the TM_{01} mode and a fictitious wave having a velocity equal to n_1/c (see the right-hand ordinate of Fig. 5.17). It is shown in Fig. 5.17 that the delay difference is reduced remarkably (to about 1 ns/km for $\Delta = 0.01$) at $\alpha = 2$; this is a well-known fact.

The difference between the rigorous value of D and that computed from the scalar wave approximation is shown in Fig. 5.18 for $\Delta = 0.1, 0.01$, and 0.001 as functions of α. When $\Delta = 0.01$, the error is approximately 0.01 ns/km (1% in D) for $\alpha = 2$, and 0.8 ns/km (6% in D) for $\alpha = 10$. Note that the percentage error in the delay difference is much larger than that in the propagation constant.

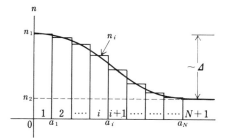

Fig. 5.19   Approximation of the refractive-index profile by a staircase function.

5.9   Analysis of Nonuniform-Core Fibers by Staircase-Approximation Method

5.9.1   Introduction

In the staircase-approximation method, the refractive-index profile is approximated by an appropriate staircase function as shown in Fig. 5.19. The wave equation (either scalar or vectorial) is solved in each stratified layer, and the solutions are then connected at the cylindrical boundaries between these layers to obtain the proper equation representing the propagation characteristics.

The staircase-approximation method has been described by Clarricoats and Chang [19], Dil and Blok [20], and Tanaka and Suematsu [21, 22]. In this section, a vectorial wave analysis using the staircase-approximation method will be described following Tanaka and Suematsu [22].

5.9.2   Basic Equations

In each stratified layer shown in Fig. 5.19, the refractive index is assumed to be constant. Therefore, from Maxwell's equations or from Eqs. (4.17)–(4.22), the following simple wave equations can be derived for the electric field components in each year:

$$\left\{ \frac{\partial^2}{\partial r^2} + \frac{1}{r}\frac{\partial}{\partial r} + \frac{\partial^2}{\partial z^2} - \frac{1}{r^2}\left(1 - \frac{\partial^2}{\partial\theta^2}\right) \right\} E_r - \frac{1}{r^2}\frac{\partial}{\partial\theta} E_\theta = -\omega^2\varepsilon\mu E_r, \quad (5.264)$$

$$\left\{ \frac{\partial^2}{\partial r^2} + \frac{1}{r}\frac{\partial}{\partial r} + \frac{\partial^2}{\partial z^2} - \frac{1}{r^2}\left(1 - \frac{\partial^2}{\partial\theta^2}\right) \right\} E_\theta + \frac{1}{r^2}\frac{\partial}{\partial\theta} E_r = -\omega^2\varepsilon\mu E_\theta, \quad (5.265)$$

$$\left\{ \frac{\partial^2}{\partial r^2} + \frac{1}{r}\frac{\partial}{\partial r} + \frac{\partial^2}{\partial z^2} + \frac{1}{r^2}\frac{\partial^2}{\partial\theta^2} \right\} E_z = -\omega^2\varepsilon\mu E_z. \quad (5.266)$$

For magnetic field components, similar relations are derived.

We first consider the transverse electric field components. Since E_r and E_θ are coupled to each other in Eqs. (5.264) and (5.265), we use, instead of E_r and E_θ, two circularly polarized (rotating) components

$$E^- = E_r - jE_\theta, \tag{5.267}$$

$$E^+ = E_r + jE_\theta. \tag{5.268}$$

By using $\partial/\partial z = -j\beta$, $\partial/\partial\theta = jn$, and the preceding new variables, we can simplify Eqs. (5.264) and (5.265) to

$$\frac{1}{r}\frac{d}{dr}\left(r\frac{dE^\pm}{dr}\right) + \left[k^2 - \beta^2 - \frac{(n \pm 1)^2}{r^2}\right]E^\pm = 0 \tag{5.269}$$

which is identical to Eq. (5.23) except that the refractive index is now assumed to be constant.

For the axial field components, we first define a parameter having the dimension of impedance,

$$Z_0 = \omega\mu/\beta, \tag{5.270}$$

and next, using Z_0, new variables associated with both the axial electric and magnetic fields,

$$\Phi_z^- = jE_z - Z_0 H_z, \tag{5.271}$$

$$\Phi_z^+ = -jE_z + Z_0 H_z. \tag{5.272}$$

From Eq. (5.266) and a corresponding expression for the axial magnetic field, we obtain

$$\beta\Phi_z^\pm = \left(\frac{d}{dr} \pm \frac{n \pm 1}{r}\right)E^\pm. \tag{5.273}$$

On the other hand, for transverse magnetic field components, if we define and use

$$H^- = H_r - jH_\theta, \tag{2.274}$$

$$H^+ = H_r + jH_\theta, \tag{5.275}$$

these are, using Φ_z^- and Φ_z^+,

$$(k^2 - \beta^2)H^\mp = \frac{\pm j}{2\omega\mu}\left(\frac{d}{dr} \pm \frac{n}{r}\right)[(k^2 \pm \beta^2)\Phi_z^- + (k^2 \mp \beta^2)\Phi_z^+]. \tag{5.276}$$

5.9.3   Propagation and Boundary Matrices

Using the four variables E^-, E^+, Φ_z^-, and Φ_z^+, we can express the field in a stratified layer and its change at a boundary by simple matrix equations.

First, electromagnetic fields at the inner surface of the ith cylindrical layer $(r = r_1 = a_i + 0)$ and at its outer surface $(r = r_2 = a_{i+1} - 0)$ are related as

$$
\begin{bmatrix} \Phi_z^- \\ E^- \\ \Phi_z^+ \\ E^+ \end{bmatrix}_{r=r_2} = \begin{bmatrix} p_{11} & p_{12} & 0 & 0 \\ p_{21} & p_{22} & 0 & 0 \\ \hline 0 & 0 & p_{33} & p_{34} \\ 0 & 0 & p_{43} & p_{44} \end{bmatrix} \begin{bmatrix} \Phi_z^- \\ E^- \\ \Phi_z^+ \\ E^+ \end{bmatrix}_{r=r_1} = [P_i] \begin{bmatrix} \Phi_z^- \\ E^- \\ \Phi_z^+ \\ E^+ \end{bmatrix}_{r=r_1} . \tag{5.277}
$$

In this expression, the matrix elements are given as the transcendental functions of r_1 and r_2 which are different for two cases $\beta < kn_i$ and $\beta > kn_i$, where n_i denotes the refractive index in the ith layer. These matrix elements are shown in Appendix 5A.12. The matrix P_i is called the propagation matrix of the ith layer.

Next, the change of the electromagnetic field at the boundary between the ith and $(i + 1)$th layers $(r = a_i)$ is expressed as

$$
\begin{bmatrix} \Phi_z^- \\ E^- \\ \Phi_z^+ \\ E^+ \end{bmatrix}_{r=a_i+0} = \begin{bmatrix} 1 & 0 & 0 & 0 \\ 0 & 1+\Delta_i & 0 & \Delta_i \\ 0 & 0 & 1 & 0 \\ 0 & \Delta_i & 0 & 1+\Delta_i \end{bmatrix} \begin{bmatrix} \Phi_z^- \\ E^- \\ \Phi_z^+ \\ E^+ \end{bmatrix}_{r=a_i-0} = [T_i] \begin{bmatrix} \Phi_z^- \\ E^- \\ \Phi_z^+ \\ E^+ \end{bmatrix}_{r=a_i-0} ,
$$
$$\tag{5.278}$$

where

$$
\Delta_i = \tfrac{1}{2}[(n_{i-1}^2/n_i^2) - 1]. \tag{5.279}
$$

The matrix T_i is called the boundary matrix.

5.9.4   Dispersion Equation

Using P_i and T_i for each layer, we can relate electromagnetic fields at the outer surface of the first later $(r = a_1 - 0)$ and at the inner surface of the $(N + 1)$th layer (cladding) by

$$
\begin{bmatrix} \Phi_z^- \\ E^- \\ \Phi_z^+ \\ E^+ \end{bmatrix}_{r=a_N+0} = [F_{ij}] \begin{bmatrix} \Phi_z^- \\ E^- \\ \Phi_z^+ \\ E^+ \end{bmatrix}_{r=a_1-0} , \tag{5.280}
$$

where

$$
[F_{ij}] = T_N(P_{N-1}T_{N-1}) \cdots (P_1 T_1). \tag{5.281}
$$

On the other hand, since the field should not diverge both in the first layer $(r < a_1)$ and in the cladding $(a_N < r)$, the fields in these two layers are simply

$$
\begin{bmatrix} \Phi_z^- \\ E^- \\ \Phi_z^+ \\ E^+ \end{bmatrix} = \begin{bmatrix} J_n(ur)/\beta & 0 \\ -J_{n-1}(ur)/u & 0 \\ 0 & J_n(ur)/\beta \\ 0 & J_{n+1}(ur)/u \end{bmatrix} \begin{bmatrix} A_1 \\ A_2 \end{bmatrix} \qquad \text{(for} \quad r < a_1), \quad (5.282)
$$

$$
\begin{bmatrix} \Phi_z^- \\ E^- \\ \Phi_z^+ \\ E^+ \end{bmatrix} = \begin{bmatrix} K_n(wr)/\beta & 0 \\ -K_{n-1}(wr)/w & 0 \\ 0 & K_n(wr)/\beta \\ 0 & -K_{n+1}(wr)/w \end{bmatrix} \begin{bmatrix} A_3 \\ A_4 \end{bmatrix} \qquad \text{(for} \quad a_N < r), \quad (5.283)
$$

where A_1, A_2, A_3, and A_4 are constants, and

$$
u = \sqrt{k^2 n_1^2 - \beta^2}, \tag{5.284}
$$

$$
w = \sqrt{\beta^2 - k^2 n_{\sqrt{N+1}}^2}, \tag{5.285}
$$

n_1 and $n_{\sqrt{N+1}}$ denoting the refractive indices in the first layer and the $(N + 1)$th layer (cladding).

Substituting Eqs. (5.281)–(5.283) into (5.280), we obtain an equation of the form

$$
\text{func}(\beta, \omega) = 0 \tag{5.286}
$$

which is the dispersion equation expressing the propagation characteristics of a nonuniform-core fiber.

5.9.5   Approximations

As the number of layers becomes infinite, the stepped dielectric profile approaches that of actual fiber profiles. Thus, computation of the dispersion relation will produce results that differ from actual values when a finite number of layers is used. However, because computation time is lengthy for large layer numbers, we must compromise between computer time and accuracy.

A possible method of reducing the computer time in the present case is to assume $\Delta_i \ll 1$ and approximate T_i as

$$
T_i' = \begin{bmatrix} 1 & 0 & 0 & 0 \\ 0 & 1 + \Delta_i & 0 & 0 \\ \hline 0 & 0 & 1 & 0 \\ 0 & 0 & 0 & 1 + \Delta_i \end{bmatrix}, \tag{5.287}
$$

or even as

$$T_i'' = \begin{bmatrix} 1 & 0 & | & 0 & 0 \\ 0 & 1 & | & 0 & 0 \\ \hline 0 & 0 & | & 1 & 0 \\ 0 & 0 & | & 0 & 1 \end{bmatrix}. \tag{5.288}$$

In Eq. (5.287), the Δ_i in diagonal elements are preserved whereas those in off-diagonals are neglected. We do this because when a product of many boundary matrices are made as in Eq. (5.281), the Δ_i in off-diagonal elements become less significant, whereas the effect of those in diagonal elements remain in the form of a product of $(1 + \Delta_i)$s.

We note here that P_i defined in Eq. (5.277) can be divided into two 2×2 matrices. Therefore, if T_i can also be divided into 2×2 matrices as in Eq. (5.287) or in Eq. (5.288), then Eq. (5.280) is divided into two independent equations

$$\begin{bmatrix} \Phi_z^- \\ E^- \end{bmatrix}_{r=a_N+0} = [F_{ij}^-] \begin{bmatrix} \Phi_z^- \\ E^- \end{bmatrix}_{r=a_1-0}, \tag{5.289}$$

$$\begin{bmatrix} \Phi_z^+ \\ E^+ \end{bmatrix}_{r=a_N+0} = [F_{ij}^+] \begin{bmatrix} \Phi_z^+ \\ E^+ \end{bmatrix}_{r=a_1-0}, \tag{5.290}$$

leading to two independent dispersion equations and thus appreciably reducing the computer time.

Comparison of Eq. (5.269) with Eqs. (5.23) and (5.78) shows that solutions derived from Eq. (5.289) correspond to HE modes, whereas those derived from Eq. (5.290) correspond to EH, TE, and TM modes.

In Refs. [21] and [22], the propagation constant of the HE_{11} mode propagating in a structure having the index profile shown in Fig. 5.10b is computed by various approximations:

β_4 exact solution based on the 4×4 matrix equation,
β_2 solution based on T_i',
β_2' solution based on T_i'',
Γ solution based on the scalar wave approximation.

It is first found that when the relative index difference between the core and cladding is less than 5%, the difference between β_2 and β_4 is more than two orders of magnitude smaller than the difference between β_2' and β_4. Hence, it is concluded that β_2 is sufficiently accurate and for all practical purposes can be regarded as the exact solution.

Differences between β_2, β_2', and Γ for the HE_{11} mode are shown in Fig. 5.20 as functions of the relative index difference Δ for the case $N = 30$

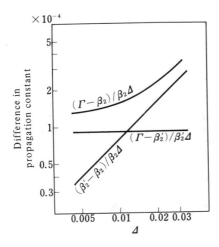

Fig. 5.20 Differences between β_2, β_2', and Γ for HE_{11} mode as functions of the relative index difference Δ (after Tanaka and Suematsu [21]).

and $v = 15$. It is found that $(\Gamma - \beta_2')/\beta_2'\Delta$ is almost independent of Δ, whereas $(\beta_2' - \beta_2)/\beta_2\Delta$ is almost proportional to Δ. From this result and the fact that the index gradient is neglected in Eq. (5.288), it is presumed in Refs. [21] and [22] that the approximation in T_i'' [Eq. (5.288)] corresponds to the scalar wave approximation.

5.10 Summary

Propagation characteristics of nonuniform-core fibers have been discussed using various wave theories. It has been shown by the WKB and power-series expansion analyses that an index profile having a quadratic core and a valley inside the core–cladding boundary is desirable for reducing the multimode dispersion. The finite element method and the staircase-approximation method have been described mainly to reveal the relation between the vectorial wave (rigorous) analysis and the scalar wave analysis.

References

1. C. N. Kurtz and W. Streifer, Guided waves in inhomogeneous focusing media—Part I: Formulation, solution for quadratic inhomogeneity, *IEEE Trans. Microwave Theory Tech.* **MTT-17**, No. 1, 11–15 (1969).
2. After the original manuscript was written, a paper appeared in which the WKB analysis is improved to make it applicable to close-to-cutoff modes: R. Olshansky, Effect of the cladding on pulse broadening in graded-index optical waveguides, *Appl. Opt.* **16**, No. 8, 2171–2174 (1977). See also, K. Oyamada and T. Okoshi, High-accuracy WKB analysis

of α-power graded-core fibers, *IEEE Trans. Microwave Theory Tech.* **MTT-28**, No. 8, 839–845 (1980).

3. D. Gloge and E. A. J. Marcatili, Impulse response of fibers with ring-shaped parabolic index distribution, *Bell Syst. Tech. J.* **52**, No. 7, 1161–1168 (1973).

4. K. Petermann, The mode attenuation in general graded-core multimode fibers, *Arch. Elektronik Ubertragungstechnik* **29**, No. 7–8, 345–348 (1975).

5. L. I. Schiff, "Quantum Mechanics." McGraw-Hill, New York, 1955.

6. W. J. Stewart, End launching of, and emission from, leaky modes in graded fibers, *Electron. Lett.* **11**, No. 21, 516–518 (1975).

7. M. Abramowitz and J. A. Stegun, "Handbook of Mathematical Functions," pp. 446–452. Dover, New York, 1965.

8. D. Gloge and E. A. J. Marcatili, Multimode theory of graded-core fibers, *Bell Syst. Tech. J.* **52**, No. 9, 1563–1578 (1973).

9. After the original manuscript was written, a paper appeared in which the y-values are shown for various dopants: H. M. Presby and I. P. Kaminov, Binary silica optical fibers: Refractive index and profile dispersion measurements, *Appl. Opt.* **15**, No. 12, 3029–3036 (1976).

10. T. Okoshi and K. Okamoto, Analysis of wave propagation in inhomogeneous optical fibers using a variational method, *IEEE Trans. Microwave Theory Tech.* **MTT-22**, No. 11, 938–945 (1974).

11. K. Okamoto and T. Okoshi, Analysis of wave propagation in optical fibers having core with α-power refractive index distribution and uniform cladding, *IEEE Trans. Microwave Theory Tech.* **MTT-24**, No. 7, 416–421 (1976).

12. K. Okamoto, T. Okoshi, and K. Hotate, A closed-form approximate dispersion formula for α-power graded-core fibers, *Fiber Integrated Opt.* **2**, No. 2, 127–142 (1979).

13. K. Oyamada and T. Okoshi, High-accuracy numerical data on propagation characteristics of α-power graded-core fibers, *IEEE Trans. Microwave Theory Tech.* **MTT-28**, No. 10, 1113–1118 (1980).

14. W. A. Gambling, D. N. Payne, and H. Matsumura, Cutoff frequency in radially inhomogeneous single-mode fiber, *Electron. Lett.* **13**, No. 5, 139–140 (1977).

15. K. Okamoto and T. Okoshi, Computer-aided synthesis of the optimum refractive index profile for a multimode fiber, *IEEE Trans. Microwave Theory Tech.* **MTT-25**, No. 3, 213–221 (1977).

16. K. Okamoto and T. Okoshi, Vectorial wave analysis of inhomogeneous optical fibers using finite element method, *IEEE Trans. Microwave Theory Tech.* **MTT-26**, No. 2, 109–114 (1978).

17. O. C. Zienkiewicz and Y. K. Cheung, "The Finite-Element Method in Continuous Structural Mechanics," McGraw Hill, New York, 1967.

18. R. Yamada and Y. Inaba, Guided waves along graded-index dielectric rod, *IEEE Trans. Microwave Theory Tech.* **MTT-22**, No. 8, 813–814 (1974).

19. P. J. B. Clarricoats and K. B. Chan, Electromagnetic wave propagation along radially inhomogeneous dielectric cylinders, *Electron. Lett.* **6**, No. 22, 694–695 (1970).

20. J. G. Dil and H. Blok, Propagation of electromagnetic surface waves in a radially inhomogeneous optical waveguide, *Opto-Electronics (London)* **5**, 415–428 (1973).

21. T. Tanaka and Y. Suematsu, Matrix analysis of graded-index cylindrical fiber with small index difference (in Japanese), Paper of Technical Group, IECE Japan, No. OQE75-26 (June 1975).

22. T. Tanaka and Y. Suematsu, An exact analysis of cylindrical fiber with index distribution by matrix method and its application to focusing fiber" *Trans. Inst. Electronics Comm. Eng. Jpn., Sect. E* **E59**, No. 11, 1–8 (1976).

6 | Classification and Comparison of Various Analysis Methods

Various methods of analysis of the propagation characteristics of modes in optical fibers are classified and compared. The relation between ray theory and wave theory is discussed using a slab model.

6.1 Introduction

Propagation characteristics of optical fibers using ray theory have been described in Chapter 3, and various other wave theories have been presented in Chapter 4 (for uniform-core fibers) and Chapter 5 (for nonuniform-core fibers). Ray theory is applicable to both uniform- and nonuniform-core fibers, and its use gives a relatively simple physical picture of propagation in optical fibers. However, ray theory is based on crude approximations and can be applied only to relatively large core fibers.

The wave theories can be classified, via the crudity of the approximations on which they are based, into three major groups: (1) the WKB analysis, (2) the scalar wave numerical analysis and (3) the vectorial wave analysis. The accuracy of the vectorial wave analysis is best and can be taken to any desired level by simply improving the precision of the numerical computation.

Various wave theories will be briefly described, classified, and compared according to their advantages and disadvantages.

The relation between the ray and wave theories will be discussed for a slab model. The WKB analysis will mainly be considered as the representative of the wave theories, because it is "closest" to the ray theory, and hence the comparison is relatively easy. The relation between concepts of the mode and ray, physical implication of the turning point in the WKB analysis, and the relation between group velocity and ray velocity will be clarified.

6.2   Various Wave Theories

6.2.1   Uniform- and Nonuniform-Core Fiber Theories

The major difference between the theories for uniform- and nonuniform-core fibers is that in the former $\nabla\varepsilon = 0$ in the core. As a result, fields in uniform-core fibers can be rigorously solved by scalar wave analysis, whereas nonuniform-core fiber fields can be rigorously solved only by vectorial wave analysis.

Only a limited number of theories have been applied to uniform-core fibers. The WKB analysis (Section 5.3) and the rigorous field analysis (Section 4.3) are sufficient to understand key features of the propagation characteristics. [The conventional WKB analysis is not applicable (see the footnote in Section 5.2.8) to fibers in which only a few modes propagate.] The field analysis based directly on Maxwell's equations as described in Section 4.3 is entirely rigorous and can be applied to single- as well as multimode fibers [1, 2].

6.2.2   Wave Theories for Nonuniform-Core Fibers

Wave theories for nonuniform-core fibers are classified as scalar wave and vectorial wave analyses, as described in Section 5.2.2. (Interesting discussions on the relation between these theories will also be found in Kurtz [3] and Yip and Nemoto [4].) As will be seen, some analysis methods can be used in both the scalar wave and vectorial wave analyses. In the following classification, these methods are included in the vectorial wave analyses. Important contributions to each method are cited as references.

A. Scalar Wave Analyses

WKB Method   For the formulation, see Section 5.3. The features are described in Section 6.2.1 [5–7].

Variational Method—Use of Variational Expression for Propagation Constant   The propagation constant is determined from its variational expression [8] by using Rayleigh–Ritz method [9] or perturbation method [10, 11].

Variational Method—Rayleigh–Ritz Method Analysis   The scalar wave equation is translated into a variational problem, and the proper equation is derived from its stationary condition. The required computer time is relatively short. See Section 5.5 [12, 13].

Power-Series Method   Both the permittivity profile and field distribution are expressed by power series, and the coefficients for the field are determined term by term. This method is useful for cases when the permittivity profile is expressed by relatively simple (short) power series [14]. See also Section 5.6.

Analytical Solution   For quadratic-index-profile fibers (without uniform cladding), the solution of the scalar wave equation can be obtained in an analytic form [15, 16]. This is not applicable, however, to those profiles having a uniform cladding.

Approximation of Vectorial Wave Analyses   As seen in Sections 5.8 and 5.9 for the finite element method and staircase-approximation method, respectively, any vectorial wave analysis can be approximated to produce a scalar wave analysis.

B. Vectorial Wave Analyses

Finite Element Method   This has been described in Section 5.8. The finite element method is suited to the analysis of optical fibers because the piecewise-linear trace of the "graded" index profile gives a relatively good approximation even when the sampling point number is small [17].

Staircase-Approximation Method   This has been described in Section 5.9. The advantage of this method is that the physical picture behind the calculation is relatively clear [18–21].

Direct Integration Method   The vectorial wave equation is first transformed into a matrix-type differential equation; it is then solved by direct integration along the radial coordinate [22].

Perturbation Method   The term including $\nabla \varepsilon$ in the vectorial wave equation is regarded as a perturbation term. The unperturbed solution (scalar wave solution) is computed first and is then corrected by taking that perturbation term into account [23, 24]. This method, however, has mainly been applied only to quadratic-core fibers.

Another type of the perturbation-method analysis is the computation of the propagation characteristics for an arbitrary profile by correcting the solution for a uniform-core fiber considering the difference in the profie as the perturbation term [25].

6.2.3   Theories Giving Mode Cutoff Frequencies

There is a category of theories aiming at the computation of mode cutoff frequencies. Gambling *et al.* proposed a series-expansion method for determining the cutoff frequencies of α-power [26] and inverse-α-power [27] nonuniform-core fibers within the scalar wave approximation. Arnold gave an asymptotic expression for mode cutoff frequencies in a quadratic-core fiber having uniform cladding [28]. Hotate and Okoshi proposed an approximate, relatively simple formula giving cutoff frequencies for arbitrary profiles [29, 30].

Equations giving the cutoff frequencies derived by the series-expansion method [26] are described in Section 5.6.4. A table of exact normalized cutoff frequencies [31] will be found in Appendix 5A.9.

6.3   Relation between Wave and Ray Theories

6.3.1   Description of the Problem

As described in Section 6.1, among various wave theory analyses, the WKB method is the closest to ray theory. The relation between the two theories is considered.

The ray trajectory is first derived from the mode concept and is shown to coincide with the trajectory computed by ray theory. Next, the relation between the propagation constants of adjacent modes and ray undulation wavelengths of that mode group is derived. Finally, it is shown that the axial velocity of the ray derived from the mode concept is equal to the average axial velocity of the meridional ray derived in Section 3.3.3.

We consider a ray propagating in a two-dimensional medium having a refractive-index distribution $n(x)$ as shown in Fig. 6.1b. It is assumed that $n(x)$ is symmetrical with respect to $x = 0$, and the propagating region (where n is elevated) is sufficiently wide so that a large number of modes can propagate. We express $n(x)$ as

$$n(x) = n_1\{1 - \tfrac{1}{2}f(x)\} \tag{6.1}$$

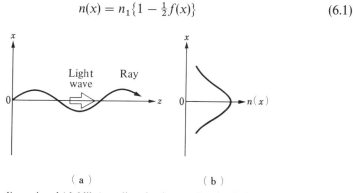

(a)   (b)

Fig. 6.1   Two-dimensional (slablike) medium having a refractive-index distribution $n(x)$.

and assume that

$$0 \leqq f(x) \ll 1, \tag{6.2}$$

$$d^2f/dx^2 \geqq 0, \tag{6.3}$$

$$f(0) = 0. \tag{6.4}$$

6.3.2 Wave Equation and Its WKB Solution

When the angular frequency of light, the propagation constant in the z direction, and the field distribution of the light are denoted by ω, β, and $\psi(x) \exp[j(\omega t - \beta z)]$, respectively, a differential equation for determining $\psi(x)$ can be written as

$$(d^2\psi/dx^2) + [\beta_1^2\{1 - f(x)\} - \beta^2]\psi = 0, \tag{6.5}$$

where

$$\beta_1^2 = \omega^2 \varepsilon_0 \mu_0 n_1^2 \tag{6.6}$$

is the propagation constant of a plane wave in a medium having refractive index n_1. Equation (6.5) is essentially identical to Eq. (5.103) and can also be derived directly from the wave equation in Cartesian coordinates (e.g., see [32]). Letting

$$1 - \beta^2/\beta_1^2 = \xi, \tag{6.7}$$

we can rewrite Eq. (6.5) as

$$(d^2\psi/dx^2) + \beta_1^2\{\xi - f(x)\}\psi = 0. \tag{6.8}$$

If the value of ξ corresponding to the proper value of Eq. (6.8) is denoted by ξ_m, the propagation constant of the mth mode β_m is expressed, from Eq. (6.7), as

$$\beta_m \simeq (1 - \xi_m/2)\beta_1. \tag{6.9}$$

If we write further

$$p(x) = \xi - f(x), \tag{6.10}$$

then Eq. (6.8) is rewritten to give

$$(d^2\psi/dx^2) + \beta_1^2 p(x)\psi = 0. \tag{6.11}$$

The notations ξ and ξ_m will be used in the following discussions.

Equation (6.11) has a form identical to Eq. (5.103); hence the former can also be solved by the WKB method. First, from Eq. (6.10), the turning points

$x = \pm A$ are determined by

$$p(x) = \xi - f(x) = 0. \tag{6.12}$$

Equation (6.11) has an oscillatory solution in the range $|x| < A$ but an exponentially decaying solution in the range $|x| > A$.

The approximate WKB solution of Eq. (6.11) in the range $|x| < A$ is [33] [see also Eq. (5.117)]

$$\psi(x) \propto p^{-1/4} \sin\left\{ \beta_1 \int_0^x \sqrt{p}\, dx + a \right\}, \tag{6.13}$$

where a is a constant. We note here that $n(x)$ is symmetrical with respect to $x = 0$ as shown in Fig. 6.1b. Hence, $\psi(x)$ must either be symmetrical or antisymmetrical with respect to $x = 0$. Therefore, a must be expressed as

$$a = (\pi/2)k, \tag{6.14}$$

where k is an integer. On the other hand, as described in Section 5.3.5, consideration of the continuity condition at the turning point leads to

$$\psi(x) \propto p^{-1/4} \sin\left(-\beta_1 \int_A^x \sqrt{p(x)}\, dx + \tfrac{1}{4}\pi \right). \tag{6.15}$$

From Eqs. (6.13) and (6.15), we obtain

$$\beta_1 \int_0^A \sqrt{p}\, dx + (\tfrac{1}{2}k + 1)\pi = -\tfrac{1}{4}\pi - 2l\pi, \tag{6.16}$$

where l is an integer. This leads further to

$$\beta_1 \int_{-A}^A \sqrt{p}\, dx = (m - \tfrac{1}{2})\pi \tag{6.17}$$

which is the proper equation corresponding to (5.139). In Eq. (6.17), m again denotes an integer; this equation gives the propagation constant of the mth mode β_m.

6.3.3   The Ray Equation and Its Solution

Next, let us calculate the trajectory, using geometrical optics, of a ray traveling in the medium depicted in Fig. 6.1. As shown in Section 2.8, the ray trajectory in an inhomogeneous medium is a solution of the ray equation [Eq. (2.68)]

$$\frac{d}{ds}\left(n\frac{d\mathbf{r}}{ds} \right) = \nabla n, \tag{6.18}$$

where s and $\mathbf{r}$ denote the curved coordinate along the ray and the position vector, respectively. If the paraxial approximation is made so that the ray travels almost along the z axis, we may write

$$ds^2 = dz^2 + dx^2 \simeq dz^2, \tag{6.19}$$

In the present case, the refractive index varies only in the x direction. Hence the x component of Eq. (6.18) becomes

$$\frac{d^2x}{dz^2} = \frac{1}{n}\frac{dn}{dx} \simeq -\frac{1}{2}\frac{df}{dx}. \tag{6.20}$$

We note here that for a specific ray, from Eq. (6.10),

$$p(x) + f(x) = \xi = \text{const.} \tag{6.21}$$

Therefore,

$$\frac{1}{2}\frac{d}{dx}(p + f) = 0, \tag{6.22}$$

and from Eqs. (6.20) and (6.22), we obtain

$$dx/dz = \sqrt{p}. \tag{6.23}$$

This equation shows that $p^{1/2}$ gives the slope of the ray with respect to the z axis.

Substituting Eq. (6.21) into (6.23) and integrating, we obtain [33]

$$z = \int \frac{dx}{\sqrt{\xi - f(x)}} + \text{const.} \tag{6.24}$$

This equation gives the trajectory of a paraxial ray. Combining Eqs. (6.4), (6.21), and (6.23), we find that $\xi^{1/2}$ gives the slope of the ray with respect to the z axis at the point where the ray crosses the axis.

6.3.4 Relation between Mode and Ray Concepts

Now we are ready to consider the relation between the concept of the mode and that of the ray. We first try to derive the trajectory of a ray on the basis of the wave theory (actually WKB analysis). It is assumed that the ray corresponding to the mth mode is given as the locus of points where phases of modes adjacent to the mth mode coincide with and enhance each other. It will be found in the following that such an assumption leads to a good correspondence between the wave and ray theories, and also to a good physical picture of the turning point [34].

We assume that a large number of modes around the mth mode, i.e., those from the $(m - \Delta m)$th through the $(m + \Delta m)$th mode, propagate simultaneously in a multimode fiber as depicted in Fig. 6.1. In such a case the field distribution is, from Eq. (6.15),

$$\psi_a(x, z, t) = \sum_{m - \Delta m}^{m + \Delta m} |A_m| p^{-1/4} \exp j\left(\omega t - \beta z \pm \beta_1 \int_0^x \sqrt{p}\, dx + \alpha_m\right). \quad (6.25)$$

In this equation, A_m and α_m are constants, and the sinusoidal function in Eq. (6.15) is replaced by an exponential function; hence, ψ_a is now complex. As is evident from the relation of the polarity with respect to the temporal term $j\omega t$, the positive and negative signs in Eq. (6.25) correspond to the "inward" wave (approaching the z axis) and the "outward" wave (leaving the z axis), respectively. Since we are now dealing with a large number of modes, we can take m to be a continuous quantity. The phase of the mth mode is then expressed as

$$\Phi(m) = \omega t - \beta z \pm \beta_1 \int_0^x \sqrt{p}\, dx + \alpha(m), \quad (6.26)$$

where $\alpha(m)$ is now a continuous function.

We assume here that the intensity of the mth mode is enhanced when the phase of the various modes around the mth mode coincide, and the "ray" corresponding to the mth mode is given as the locus of such points. The condition for such points is, from the stationary condition of Φ_m,

$$d\Phi(m)/dm = (d\xi/dm)\{d\Phi(m)/d\xi\} = 0. \quad (6.27)$$

If we consider the outward wave, Eqs. (6.9), (6.10), (6.26), and (6.27) lead to

$$\frac{d\xi}{dm}\left(z - \int_0^x \frac{1}{\sqrt{p}}\, dx + \frac{2}{\beta_1}\frac{d\alpha}{d\xi}\right) = 0. \quad (6.28)$$

Therefore,

$$z = \int_0^x \frac{1}{\sqrt{p}}\, dx - \frac{2}{\beta_1}\frac{d\alpha}{d\xi}. \quad (6.29)$$

Next, for an initial point (x_0, z_0), we may also write

$$\omega t - \beta z_0 + \beta_1 \int_0^{x_0} \sqrt{p}\, dx + \alpha(m) = \text{const.} \quad (6.30)$$

Equation (6.29) holds also at this point;

$$z_0 = \int_0^{x_0} \frac{1}{\sqrt{p}}\, dx - \frac{2}{\beta_1}\frac{d\alpha}{d\xi}. \quad (6.31)$$

Combining Eqs. (6.29) and (6.31) and using Eq. (6.10), we obtain

$$z - z_0 = \int_{x_0}^{x} \frac{dx}{\sqrt{\xi - f(x)}}. \tag{6.32}$$

This equation gives the trajectory of the ray derived by the wave theory under the assumption just stated. Since Eqs. (6.24) and (6.32) are identical, the foregoing assumption is reasonable.

Other relations between wave and ray theories can be considered. First, using the relation between β and ξ [see Eq. (6.7)] and being aware of the physical implication of ξ described in Section 6.3.3, we find the following relation: the higher the mode order (i.e., the smaller the value of β_m), the greater the slope between the z axis and the ray corresponding to that mode.

Second, the furthest point of ray travel given from Eqs. (6.21) and (6.23), both of which are based on ray theory, satisfies

$$f(x) = \xi. \tag{6.33}$$

This agrees with Eq. (6.12) derived by WKB analysis. The term "turning point" used conventionally in WKB analysis stems from the fact that the turning point is also the place at which the trajectory of the ray corresponding to a specific mode becomes parallel to the z-axis.

6.3.5   Relation between Propagation Constant and Undulation Wavelength

Next we derive an interesting relation between the propagation constants of adjacent modes and the wavelength of the ray undulation corresponding to that mode group.

According to WKB analysis, the propagation constant β is determined by Eq. (6.17). Therefore, if we define a new function of ξ,

$$N(\xi) = (\beta_1/\pi) \int_{-A}^{A} \sqrt{p}\, dx, \tag{6.34}$$

$N(\xi)$ changes stepwise by one as the mode changes, because m in Eq. (6.17) is an integer. Therefore, the difference of β between two adjacent modes is

$$\Delta\beta = d\beta/dN = (d\beta/d\xi)/(dN/d\xi) = \pi \bigg/ \int_{-A}^{A} (1/\sqrt{p})\, dx. \tag{6.35}$$

On the other hand, on the basis of ray theory, the undulation wavelength of a ray Λ is, from Eqs. (6.24) and (6.21),

$$\Lambda = 2 \int_{-A}^{A} (1/\sqrt{p})\, dx. \tag{6.36}$$

From Eqs. (6.35) and (6.36),

$$\Delta\beta\,\Lambda = 2\pi. \qquad (6.37)$$

Thus it is found that the phase difference between two adjacent modes shifts 2π while the corresponding ray travels over one wavelength of its undulation.

6.3.6   Relation between Ray and Group Velocities

In Section 3.3.3, the average velocity of a meridional ray is computed by ray theory. In the following, we compute the average velocity of a ray corresponding to a mode as the average group velocity of the mode group surrounding that mode. It will be found that the result agrees with that of the ray theory shown in Section 3.3.3 [34].

We consider, as in Section 6.3.4, that the $(m - \Delta m)$th–$(m + \Delta m)$th modes propagate in a multimode fiber. In the present case we also consider that the angular frequency of light is distributed between $\omega - \Delta\omega$ and $\omega + \Delta\omega$. (The group velocity cannot be computed without such a frequency spread.) Then the field distribution is, as in Eq. (6.25),

$$\psi_a = \int_{\omega-\Delta\omega}^{\omega+\Delta\omega} d\omega \sum_{m-\Delta m}^{m+\Delta m} |A(\omega, m)| p^{-1/4} \exp j\left(\omega t - \beta z \pm \beta_1 \int \sqrt{p}\,dx + \alpha(\omega, m)\right).$$

$$(6.38)$$

The ray trajectory corresponding to the mth mode with angular frequency ω can be computed, as has been done in Section 6.3.4, by tracing the locus of points at which the phases of various components included in Eq. (6.38) coincide. We again consider the outward wave as in the derivation of Eq. (6.29).

If we denote the phase term of a component included in Eq. (6.38) by $\Phi(\omega, m, x, z, t)$, the ray trajectory is given by

$$\partial\Phi/\partial\omega = 0, \qquad (6.39)$$

$$\partial\Phi/\partial m = 0. \qquad (6.40)$$

If the phases of all components coincide at $t = t_0$, $x = x_0$, and $z = z_0$, then

$$\omega t_0 - \beta z_0 - \beta_1 \int_0^{x_0} \sqrt{p}\,dx + \alpha(\omega, m) = \text{const.} \qquad (6.41)$$

Therefore, first computing Eqs. (6.39) and (6.40) and then substituting $(\partial\alpha/\partial\omega)$ and $(\partial\alpha/\partial m)$ computed from Eq. (6.41) into those equations, we obtain

$$(t - t_0) - (\partial\beta/\partial\omega)(z - z_0) + \{(\partial\beta/\partial\omega)\beta$$
$$- \omega\varepsilon_0\mu_0 n^2\}(x - x_0)(\omega^2\varepsilon_0\mu_0 n^2 - \beta)^{-1/2} = 0 \qquad (6.42)$$

and

$$-(\partial\beta/\partial m)(z - z_0) + (\partial\beta/\partial m)\beta(x - x_0)(\omega^2\varepsilon_0\mu_0 n^2 - \beta^2)^{-1/2} = 0. \quad (6.43)$$

In deriving these equations, $[p(x)]^{1/2}$ is given by $n(x)$ using Eqs. (6.1), (6.6), (6.7), and (6.10), and approximations are made to facilitate the integration [34].

Equations (6.42) and (6.43) are simultaneous equations with respect to $(z - z_0)$ and $(x - x_0)$. These equations yield the ray trajectory corresponding to a mode having a propagation constant β as

$$z = z_0 + v_l \cos\theta \cdot (t - t_0), \quad (6.44)$$

$$x = x_0 + v_l \sin\theta \cdot (t - t_0), \quad (6.45)$$

where

$$v_l = (\varepsilon_0\mu_0 n^2)^{-1/2}, \quad (6.46)$$

$$\tan\theta = (\omega^2\varepsilon_0\mu_0 n^2 - \beta^2)^{1/2}/\beta. \quad (6.47)$$

Note that v_l and θ are both functions of the position (z, x). Obviously, the above four equations indicate that the ray at a position (z, x) travels with a velocity v_l determined by the refractive index at that position $n(x)$ and with a slant angle given by Eq. (6.47). The slant angle θ is maximized at the axis $(x = 0)$ and is zero at turning points $(x = A)$ where the ray turns inward.

Next we consider the axial velocity of a ray, which is given by $v_l \cos\theta$. Both v_l and $\cos\theta$ are minimized at $x = 0$ and maximized at $x = A$. Hence, the axial velocity varies along the ray path. Apparently, such a property seems to contradict the concept of group velocity in the wave theory. However, the following consideration [34] will answer this question.

We start from Eqs. (6.42) and (6.43). Equation (6.43), which is derived from the phase-coincidence condition between modes, gives an undulating trajectory around the z axis. The trajectory thus given by wave theory agrees with that given by ray theory described in Section 6.3.4. On the other hand, Eq. (6.42), which is derived from the phase-coincidence condition between different frequencies, gives a relation between x, z, and t. At a given time t, the z–x relation is expressed like an "integral symbol" as shown by solid curves in Fig. 6.2. This integral symbol, within the approximation made in deriving Eq. (6.42), translates in the z direction without changing its shape at a constant velocity equal to the group velocity $\partial\omega/\partial\beta$.

In Fig. 6.2, the intermodal phase-coincidence locus is shown by a dashed curve. If we assume that the cross points between the interfrequency and intermodal phase-coincidence loci (small circles in Fig. 6.2) give the front of the ray corresponding to a mode, and that the front moves along the dashed

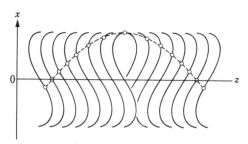

Fig. 6.2 Relation between mode and ray, —, interfrequency phase-coincidence loci; ---,
intermodal phase-coincidence locus (after Kawakami and Nishizawa [34]).

curve as shown in the figure, we can understand why ray velocity fluctuates
while the group velocity remains constant. As seen in Fig. 6.2, the "integral
sign" is inverted abruptly when the ray trajectory reaches the turning point;
the inversion corresponds to the transition from the outward wave to the
inward wave.

Finally, let us compute the average axial velocity of the ray and show
that it is identical to the group velocity. We consider that the ray front is
located at $x = z = 0$ at $t = 0$, and it arrives at $z = \Lambda$, $x = 0$ at $t = T$ after
traveling over one undulation cycle. Then from Eqs. (6.44) and (6.45), we
obtain

$$T = 2 \int_{-A}^{A} \frac{dt}{dx}\, dx = 2 \int_{-A}^{A} \frac{\omega \varepsilon_0 \mu_0 n^2}{(\omega^2 \varepsilon_0 \mu_0 n^2 - \beta^2)^{1/2}}\, dx, \qquad (6.48)$$

$$\Lambda = 2 \int_{x=-A}^{x=A} dx/\tan\theta = 2 \int_{-A}^{A} \frac{\beta}{(\omega^2 \varepsilon_0 \mu_0 n^2 - \beta^2)^{1/2}}\, dx. \qquad (6.49)$$

Therefore, the average axial velocity is

$$v_{\text{av}} = \int_{-A}^{A} \frac{\beta}{(\omega^2 \varepsilon_0 \mu_0 n^2 - \beta^2)^{1/2}}\, dx \Big/ \int_{-A}^{A} \frac{\omega \varepsilon_0 \mu_0 n^2}{(\omega^2 \varepsilon_0 \mu_0 n^2 - \beta^2)^{1/2}}\, dx. \quad (6.50)$$

On the other hand, if we write

$$I = \int_{-A}^{A} (\omega^2 \varepsilon_0 \mu_0 n^2 - \beta^2)^{1/2}\, dx, \qquad (6.51)$$

the group velocity is

$$\frac{d\omega}{d\beta} = -\frac{\partial I/\partial \beta}{\partial I/\partial \omega}$$

$$= \int_{-A}^{A} \frac{\beta}{(\omega^2 \varepsilon_0 \mu_0 n^2 - \beta^2)^{1/2}}\, dx \Big/ \int_{-A}^{A} \frac{\omega \varepsilon_0 \mu_0 n^2}{(\omega^2 \varepsilon_0 \mu_0 n^2 - \beta^2)^{1/2}}\, dx, \quad (6.52)$$

which agrees with Eq. (6.50) [34]. Note that Eqs. (6.50) and (6.52) also agree with Eq. (3.30) which gives the average axial velocity computed by ray theory.

6.4 Summary

Various methods for the analysis of propagation characteristics of optical fibers have been classified, and their advantages and disadvantages have been discussed. Various features inherent in each of the methods described in Chapter 5 have been clarified. Finally, the relation between the ray and wave theories has been discussed; correspondences between concepts in the two different theories have been elucidated.

References

1. E. Snitzer, Cylindrical dielectric waveguide modes, *J. Opt. Soc. Am.* **51**, No. 5, 491–498 (1961).
2. G. Biernson and D. J. Kinsley, Generalized plots of mode patterns in a cylindrical dielectric waveguide applied to retinal cones, *IEEE Trans. Microwave Theory Tech.* **MTT-13**, No. 5, 345–356 (1965).
3. C. N. Kurtz, Scalar and vector mode relations in gradient-index light guides, *J. Opt. Soc. Am.* **65**, No. 11, 1235–1240 (1975).
4. G. L. Yip and S. Nemoto, The relation between scalar modes in a lenslike medium and vector modes in a self-focusing optical fiber, *IEEE Trans. Microwave Theory Tech.* **MTT-23**, No. 2, 260–263 (1975).
5. W. Streifer and C. N. Kurtz, Scalar analysis of radially inhomogeneous guiding media, *J. Opt. Soc. Am.* **57**, No. 6, 779–786 (1967).
6. C. N. Kurtz and W. Streifer, Guided waves in inhomogeneous focusing media—Part II: Asymptotic solution for general weak inhomogeneity, *IEEE Trans. Microwave Theory Tech.* **MTT-17**, No. 5, 250–253 (1969).
7. D. Gloge and E. A. J. Marcatili, Multimode theory of graded-core fibers, *Bell Syst. Tech. J.* **52**, No. 9, 1563–1578 (1973).
8. K. Kurokawa, Electromagnetic waves in waveguides with wall impedance, *IRE Trans. Microwave Theory Tech.* **MTT-10**, No. 9, 314–320 (1962).
9. M. Matsuhara, Analysis of TEM modes in dielectric waveguides, by a variational method, *J. Opt. Soc. Am.* **63**, No. 12, 1514–1517 (1973).
10. H. J. Heyke and M. H. Kuhn, Dispersion characteristics of general gradient fibers, *Arch. Elektronik Ubertragungstechnik* **27**, No. 5, 235–238 (1973).
11. M. Ootaka, M. Matsuhara, and N. Kumagai, Analysis of lens-like media using the variational method (in Japanese). *Trans. IECE Jpn.* **55-B**, No. 6, 332–333 (1972).
12. T. Okoshi and K. Okamoto, Analysis of wave propagation in inhomogeneous optical fibers using a variational method, *IEEE Trans. Microwave Theory Tech.* **MTT-22**, No. 11, 938–945 (1974).

13. K. Okamoto and T. Okoshi, Analysis of wave propagation in optical fibers having core with α-power refractive index distribution and uniform cladding, *IEEE Trans. Microwave Theory Tech.* **MTT-24**, No. 7, 416–421 (1976).

14. H. Kirchhoff, Wave propagation along radially inhomogeneous glass fibers, *Arch. Elektronik Ubertragungstechnik* **27**, No. 1, 13–18 (1973).

15. C. N. Kurtz and W. Streifer, Guided waves in inhomogeneous focusing media—Part I: Formulation, solution for quadratic inhomogeneity, *IEEE Trans. Microwave Theory Tech.* **MTT-17**, No. 1, 11–15 (1969).

16. R. Yamada and Y. Inaba, Guided waves along graded index dielectric rod, *IEEE Trans. Microwave Theory Tech.* **MTT-22**, No. 8, 813–814 (1974).

17. K. Okamoto and T. Okoshi, Vectorial wave analysis of inhomogeneous optical fibers using finite element method, *IEEE Trans. Microwave Theory Tech.* **MTT-26**, No. 2, 109–114 (1978).

18. P. J. B. Clarricoats and K. B. Chan, Electromagnetic-wave propagation along radially inhomogeneous dielectric cylinders, *Electron. Lett.* **6**, No. 22, 694–695 (1970).

19. K. B. Chan and P. J. B. Clarricoats, Propagation characteristics of an optical waveguide with a diffused core boundary, *Electron. Lett.* **6**, No. 23, 748–749 (1970).

20. J. G. Dil and H. Blok, Propagation of electromagnetic surface waves in a radially inhomogeneous optical waveguide, *Opto-electronics (London)* **5**, 415–428 (1973).

21. T. Tanaka and Y. Suematsu, An exact analysis of cylindrical fiber with index distribution by matrix method and its application to focusing fiber, *Trans. Inst. Electron. Commun. Eng. Jpn.* **E59**, No. 11, 1–8 (1976).

22. G. L. Yip and Y. H. Ahmew, Propagation characteristics of radially inhomogeneous optical fiber, *Electron. Lett.* **10**, No. 4, 37–38 (1974).

23. M. Matsuhara, Analysis of electromagnetic-wave modes in inhomogeneous dielectric line and its application to lens-like media (in Japanese), *Trans. IECE Jpn.* **56-B**, No. 1, 9–13 (1973).

24. Y. Miyazaki, Vector wave analysis of dispersion in gradient fibers, *Arch. Elektronik Ubertragungstechnik* **29**, No. 5, 205–211 (1975).

25. A. W. Snyder, Mode propagation in optical waveguides, *Electron. Lett.* **6**, No. 18, 561–562 (1970).

26. W. A. Gambling, D. N. Payne, and H. Matsumura, Cutoff frequency in radially inhomogeneous single-mode fiber, *Electron. Lett.* **13**, No. 5, 139–140 (1977).

27. W. A. Gambling, D. N. Payne, and H. Matsumura, Effect of dip in the refractive index on the cutoff frequency of a single-mode fiber, *Electron. Lett.* **13**, No. 7, 174–175 (1977).

28. J. M. Arnold, Asymptotic evaluation of the normalized cutoff frequencies of an optical waveguide with quadratic index variation, *Microwaves Opt. Acoust. (IEE)* **1**, No. 6, 203–208 (1977).

29. K. Hotate and T. Okoshi, Formula giving single-mode limit of optical fiber having arbitrary refractive-index profile, *Electron. Lett.* **14**, No. 8, 246–248 (1978).

30. K. Hotate and T. Okoshi, A general formula giving cutoff frequencies of modes in an optical fiber having arbitrary refractive index profile, *Trans. IECE Jpn. Sect. E* **E62**, No. 1, 1–6 (1979).

31. K. Oyamada and T. Okoshi, High-accuracy numerical data on propagation characteristics of α-power graded-core fibers, *IEEE Trans. Microwave Theory Tech.* **MTT-28**, No. 10, 1113–1118 (1980).

32. L. I. Schiff, "Quantum Mechanics." McGraw-Hill, New York, 1955.

33. J. P. Gordon, Optics of general guiding media, *Bell Syst. Tech. J.* **45**, No. 2, 321–332 (1966).

34. S. Kawakami and J. Nishizawa, Kinetics of an optical wave packet in a lens-like medium, *J. Appl. Phys.* **38**, No. 12, 4807–4811 (1967).

7 | Optimum Refractive-Index Profile of Optical Fibers

Two problems of optical fibers concerning the optimum index profiles that minimize dispersion are presented.

The optimum-profile design for single-mode fibres is considered first. It is shown that by employing the so-called W-shaped profile in which a low-index intermediate layer is provided between core and cladding, we can realize a negative waveguide dispersion to cancel positive material dispersion at wavelengths $\lambda < 1.2$ μm.

We also consider the optimum index profile minimizing multimode dispersion, a phenomenon that limits the transmission bandwidth of thick fibers. In Sections 5.4.2, 5.4.3, and 5.7, we discussed this problem but considered only the optimum profile of the α-power fibers. Here, a more generalized approach to finding the optimum profile is described.

7.1 Introduction

The problem of finding the optimum refractive-index profile has been addressed by several researchers; e.g., Kawakami and Nishida [1, 2] (single-mode fibers) and Gloge and Marcatili [3], Okamoto and Okoshi [4, 5], and Furuya et al. [6] (multimode fibers). In all these papers, the optimum profile was pursued so that the overall dispersion was minimized, in other words, the transmission bandwidth was maximized.

For single-mode fibers, material and waveguide dispersions primarily limit the transmission bandwidth because the multimode dispersion is absent. (When the fiber is axially nonsymetrical, a kind of multimode dispersion due to the delay difference between vertically and horizontally polarized HE_{11} modes may be present. This effect, however, is not considered here.) Therefore, if we succeed in making the material and waveguide

166

dispersions cancel each other by properly shaping the profile, we can expect to have an extremely wide bandwidth.

There is another purpose for properly shaping the profile in a single-mode fiber. When the relative index difference Δ becomes large, the field is strongly confined in the core and hence the microbending loss (radiation loss due to short-wavelength fiber bending) is reduced, whereas the junction loss (loss at junction of two fibers due to lateral displacement) increases because the core radius decreases [see Eq. (4.89)] and vice versa. In short, the reductions of microbending loss and junction loss are contradictory requirements. It is expected that a compromise between these losses could be found if we properly shape the index profile. However, this second-kind problem has not yet been solved successfully.

The problem considered here first is the cancellation of material/waveguide dispersions in single-mode fibers. It is shown that in some cases the so-called W-shaped profile is effective for this cancellation.

For thick fibers, the multimode dispersion is the principal cause that limits the transmission bandwidth. Therefore, the profile should be controlled to minimize multimode dispersion. This problem has been discussed by many investigators [3, 4, 6] and also in Sections 5.4.2, 5.4.3, and 5.7, where the index profile was assumed to be the α-power of the radial coordinate. Thus, the profile obtained cannot be optimum.

An approach aimed at the true optimum is described here. It is a fully computer-oriented iterative synthesis of the optimum profile; in the course of the synthesis, the index profile is expressed by a radial-coordinate power series to assure the degree of freedom in the synthesis. It is shown that the optimum profile is a smoothed W-shaped one in which the central part is very close to a parabolic curve [5].

7.2 Optimum Refractive-Index Profile for Single-Mode Fibers

7.2.1 Cancellation of Material and Waveguide Dispersions

As stated in the Introduction, two kinds of optimization are possible for single-mode fibers. One is the cancellation of the material and waveguide dispersions, and the other is the best compromise between microbending and junction losses. Only the former problem is considered here because the latter has not yet been solved successfully.

In uniform-core single mode fibers, as seen at the left side of Fig. 4.10, the normalized waveguide dispersion $c\sigma_w$ is positive in the single-mode frequency range. On the other hand, as seen in Fig. 4.11b, the normalized material dispersion $c\sigma_m$ of ordinary silica glass is also positive when the wavelength $\lambda < 1.2$ μm. Therefore, these two dispersions can never cancel each other for a uniform core and $\lambda < 1.2$ μm.

When the core is uniform, the cancellation can be expected only in the long-wavelength range ($\lambda \gtrsim 1.2$ μm). It is noteworthy that the use of the long-wavelength range is also advantageous in that the transmission loss is reduced remarkably at wavelengths above 1.0 μm (see Fig. 2.6). Practically, the present problem at the long wavelengths is that a stable light source (laser) and an efficient detector are not yet available. However, these problems are outside the main scope of this book; only a short comment will be found in Chapter 11.

In the wavelength range of contemporary (1978–1979) practical semi-conductor laser (0.8–1.0 μm) where the material dispersion is positive, the total dispersion can be minimized only if the waveguide dispersion is negative. For this purpose, Kawakami and Nishida [1] investigated the propagation characteristics of fibers having a low-index intermediate layer between the core and cladding, and found that in W-type fibers (as they name it, because the profile is W-shaped), negative waveguide dispersion can be obtained in the single-mode region. Their theory will be briefly described.

7.2.2 Dispersion Equation

As shown in Fig. 7.1, we express the refractive-index profile of the W-type fiber as

$$n = \begin{cases} n_1 & (0 \le r \le a), & (7.1a) \\ n_1(1 - 2\delta)^{1/2} = n_I & [a < r \le (1 + h)a], & (7.1b) \\ n_1(1 - 2\Delta)^{1/2} = n_2 & [(1 + h)a < r]. & (7.1c) \end{cases}$$

In such a structure, the electromagnetic energy is confined to the core region, only if

$$kn_2 < \beta \le kn_1. \tag{7.2}$$

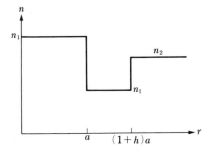

Fig. 7.1 Parameters expressing a W-type profile.

In this case, the axial components of the electric and magnetic fields are expressed as

$$E_z = AJ_n(pr)\cos n\theta \tag{7.3a}$$
$$H_z = BJ_n(pr)\sin n\theta \tag{7.3b}$$
$$[0 \leqq r \leqq a],$$

$$E_z = [CI_n(\hat{q}r) + DK_n(\hat{q}r)]\cos n\theta \tag{7.4a}$$
$$H_z = [FI_n(\hat{q}r) + GK_n(\hat{q}r)]\sin n\theta \tag{7.4b}$$
$$[a < r \leqq (1+h)a],$$

$$E_z = LK_n(qr)\cos n\theta \tag{7.5a}$$
$$H_z = MK_n(qr)\sin n\theta \tag{7.5b}$$
$$[(1+h)a < r],$$

where

$$p^2 = k^2 n_1^2 - \beta^2, \tag{7.6}$$
$$\hat{q}^2 = \beta^2 - k^2 n_I^2, \tag{7.7}$$
$$q^2 = \beta^2 - k^2 n_2^2. \tag{7.8}$$

Other field components are obtained by substituting Eqs. (7.3)–(7.5) into (5.11) and (5.12).

Next, from the condition that the tangential components of the electric and magnetic fields be continuous at $r = a$ and $r = (1+h)a$, the following matrix equation is obtained [1]:

$$
\begin{bmatrix}
n\left(\dfrac{1}{u^2} + \dfrac{1}{\hat{w}^2}\right)\hat{I}_i & n\left(\dfrac{1}{u^2} + \dfrac{1}{\hat{w}^2}\right)\hat{K}_i & \cdots \\
\dfrac{kn_1}{\beta}\hat{I}_i\left[\dfrac{J'}{uJ} + (1-2\delta)\dfrac{\hat{I}_i}{\hat{w}\hat{I}_i}\right] & \dfrac{kn_1}{\beta_1}\hat{K}_i\left[\dfrac{J'}{uJ} + (1-2\delta)\dfrac{\hat{K}_i'}{\hat{w}\hat{K}_i}\right] & \\
\dfrac{n\hat{I}_0}{(1+h)}\left(\dfrac{1}{w^2} - \dfrac{1}{\hat{w}^2}\right) & \dfrac{n\hat{K}_0}{(1+h)}\left(\dfrac{1}{w^2} - \dfrac{1}{\hat{w}^2}\right) & \\
\dfrac{kn_1}{\beta}\hat{I}_0\left[(1-2\Delta)\dfrac{K_0'}{wK_0} - (1-2\delta)\dfrac{\hat{I}_0'}{\hat{w}\hat{I}_0}\right] & \dfrac{kn_1}{\beta}\hat{K}_0\left[(1-2\Delta)\dfrac{K_0'}{wK_0} - (1-2\delta)\dfrac{\hat{K}_0'}{\hat{w}\hat{K}_0}\right] &
\end{bmatrix}
$$

$$
\begin{bmatrix}
\dfrac{kn_1}{\beta}\left(\dfrac{J'}{uJ} + \dfrac{\hat{I}_i'}{\hat{w}\hat{I}_i}\right) & \dfrac{kn_1}{\beta}\hat{K}_i\left(\dfrac{J'}{uJ} + \dfrac{\hat{K}_i'}{\hat{w}\hat{K}_i}\right) \\
n\left(\dfrac{1}{u^2} + \dfrac{1}{\hat{w}^2}\right)\hat{I}_i & n\left(\dfrac{1}{u_2} + \dfrac{1}{\hat{w}^2}\right)\hat{K}_i \\
\dfrac{kn_1}{\beta}\hat{I}_0\left(\dfrac{K_0'}{wK_0} - \dfrac{\hat{I}_0'}{\hat{w}\hat{I}_0}\right) & \dfrac{kn_1}{\beta}\hat{K}_0\left(\dfrac{K_0'}{wK_0} - \dfrac{\hat{K}_0'}{\hat{w}\hat{K}_0}\right) \\
\dfrac{n\hat{I}_0}{(1+h)}\left(\dfrac{1}{w^2} - \dfrac{1}{\hat{w}^2}\right) & \dfrac{n\hat{K}_0}{(1+h)}\left(\dfrac{1}{w^2} - \dfrac{1}{\hat{w}^2}\right)
\end{bmatrix}
\begin{bmatrix}
C \\
D \\
(\mu_0/n_1^2\varepsilon_0)^{1/2}F \\
(\mu_0/n_1^2\varepsilon_0)^{1/2}G
\end{bmatrix} = 0, \tag{7.9}
$$

where n is identical to that used in Eqs. (7.3)–(7.5), and other symbols and variables are defined as

$$u = pa, \tag{7.10}$$

$$\hat{w} = \hat{q}a, \tag{7.11}$$

$$w = qa, \tag{7.12}$$

$$J = J_n(u), \qquad\qquad J' = J'_n(u), \tag{7.13}$$

$$\hat{I}_i = I_n(\hat{w}), \qquad\qquad \hat{I}'_i = I'_n(\hat{w}), \tag{7.14}$$

$$\hat{K}_i = K_n(\hat{w}), \qquad\qquad \hat{K}'_i = K'_n(\hat{w}), \tag{7.15}$$

$$\hat{I}_0 = I_n[(1 + h)\hat{w}], \qquad \hat{I}'_0 = I'_n[(1 + h)\hat{w}], \tag{7.16}$$

$$\hat{K}_0 = K_n[(1 + h)\hat{w}], \qquad \hat{K}'_0 = K'_n[(1 + h)\hat{w}], \tag{7.17}$$

$$K_0 = K_n[(1 + h)w], \qquad K'_0 = K'_n[(1 + h)w]. \tag{7.18}$$

If we express the 4×4 matrix in Eq. (7.9) by D, a nontrival solution of Eq. (7.9) exists only if

$$\det(D) = 0. \tag{7.19}$$

This is the dispersion equation for the W-type fiber, because ω and β are included in matrix D.

7.2.3   Dispersion Characteristics

The dispersion in a single-mode fiber is given as the sum of material and waveguide dispersions. Hence, from Eqs. (4.106) and (4.107), we may write

$$\sigma = \sigma_m + \sigma_w = (\lambda^2/c)\, d^2 n/d\lambda^2 + [\omega_0 d^2\beta/d\omega^2]_{\text{waveguide}}. \tag{7.20}$$

In the preceding expression, it is assumed for simplicity that N (group index) $\simeq n$. (Refer to the last paragraph of Section 4.4.3.)

As seen in Fig. 4.10, the normalized waveguide dispersion $c\sigma_w$ in a uniform-core fiber is positive in the frequency range $v < 2.9$. This frequency limit $v = 2.9$ is higher than the single-mode limit, i.e., the cutoff frequency of the LP_{11} mode given by $v = 2.41$. Our problem is how to make σ_w negative at frequencies below the LP_{11}-mode cutoff.

Figure 7.2 shows the u–w relations for the uniform-core fiber [curves (a)] and two W-type fibers [curves (b) and (c)]. It is seen that when the refractive index of the intermediate layer is lowered (i.e., when δ becomes greater), u becomes almost constant for the variation of w [see curves (c)]. On the

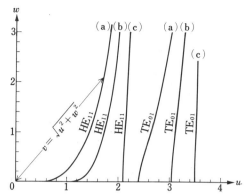

Fig. 7.2   The u–w relations for uniform-core and W-type fibers (after Kawakami and Nishida [1]): (a) uniform-core, $n_2/n_1 = 0.99$, (b) W-type, $n_I/n_1 = 0.978$, $n_2/n_1 = 0.99$, $h = 0.4142$, (c) W-type, $n_I/n_1 = 0.90$, $n_2/n_1 = 0.99$, $h = 0.4142$.

other hand, Eqs. (7.6) and (7.10) lead to

$$\beta = [((n_1/c)\omega)^2 - (u/a)^2]^{1/2}. \tag{7.21}$$

Hence, if $u/a \simeq$ const, we find by differentiating the foregoing equation twice that

$$[d^2\beta/d\omega^2] < 0. \tag{7.22}$$

This means that by employing the W-shaped profile, we can make the waveguide dispersion negative in the single-mode region, and thus offset the material dispersion even at wavelengths of $\lambda < 1.2\ \mu$m.

To assure wide freedom in such a cancellation scheme, it is desirable to know beforehand under what conditions a negatively large waveguide dispersion can be obtained. Kawakami and Nishida [1] concluded that it is desirable to make the refractive index of the intermediate layer n_I as low as possible ($n_I/n_1 \lesssim 0.8$), and also make the cladding index n_2 fairly low within the weakly guiding range ($n_2/n_1 \lesssim 0.99$). To date, however, net cancellation between material and waveguide dispersions has not actually been achieved because the deep index valley (as low as $n_I/n_1 \lesssim 0.8$) is difficult to realize with present materials.

7.3   Optimum Refractive-Index Profile for Multimode Fibers

7.3.1   Outline of the Iterative Synthesis Process

The computer-oriented, iterative synthesis of the optimum refractive-index profile for multimode fiber will be described [5].

We first express the refractive index in the core by a power series in terms of r, and use the variational method to obtain the delay time of each propagation mode. Next we compute the variance of the group delay. Then we modify the refractive-index profile so as to decrease the group delay toward its minimum. We repeat such a process of analysis, evaluation, and modification until we obtain the optimum index profile with which the group delay is minimized.

The symbols defined and used in Chapters 4 and 5 will again be used, sometimes without detailed explanations. The index distribution is assumed to be axially symmetric. The following quantities are assumed constrained during the optimization process:

(1)   wavelength λ (or angular frequency ω) of light,
(2)   maximum refractive index in the core n_1 and cladding index n_2,
(3)   number of propagating LP modes N.

Condition (2) implies that the relative index difference Δ defined in Eq. (3.4) is also constant during the optimization process. We assume also that all LP modes are equally excited and are subject to an equal transmission loss.

The refractive-index profile is expressed throughout the entire optimization process by a power series

$$n^2(r) = n_1^2[1 - 2\Delta g(r)] \qquad (0 \leq r \leq a), \qquad (7.23)$$

$$g(r) = \sum_{p=1}^{n} \kappa_p[(r/a)^{2p} - (r_0/a)^{2p}], \qquad (7.24)$$

where r_0 denotes the radial coordinate at which $n(r)$ is maximized as shown in Fig. 7.3, and the κ_p are the parameters representing the profile. Our task is to obtain the optimum set κ_p.

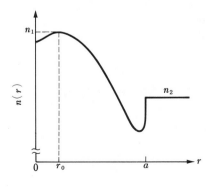

Fig. 7.3 Representation of the refractive-index profile used in the synthesis process.

7.3.2 Analysis of Propagation Characteristics by the Rayleigh–Ritz Method

The first step in the optimum-profile synthesis is the analysis of the propagation characteristics of a fiber having the starting profile. (Similar analyses are repeated afterwards for each modified profile.) For this purpose we use the Rayleigh–Ritz method described in Section 5.5. The details of the method will be omitted here; we should note that when the index profile is expressed by a power series of r as in Eqs. (7.23) and (7.24), the coefficient A_{mkl} defined in Eq. (5.194) becomes

$$A_{mkl} = \sum_{p=1}^{n} \kappa_p D_{pmkl}, \tag{7.25}$$

where

$$D_{pmkl} = 2 \int_0^1 x^{2p} \frac{J_m(\lambda_k x) J_m(\lambda_l x)}{J_m(\lambda_k) J_m(\lambda_l)} \, x \, dx - \left(\frac{r_0}{a}\right)^{2p} [1 + (\Phi_\beta^2 - m^2)/\lambda_k^2] \delta_{kl}. \tag{7.26}$$

All the symbols used in the foregoing equations are explained in Section 5.5.

Once the A_{mkl} are determined, for each mode we can compute the propagation constant β by using Eq. (5.197) and the delay time t by using Eq. (5.203).

7.3.3 Measure of Multimode Dispersion

To determine the optimum refractive-index profile, we need an object function which is a measure of multimode dispersion. In the present synthesis, we use $Q(N)$ defined by Eq. (5.231).

We consider that the optimum refractive-index profile of an optical fiber in which N modes propagate is determined by the condition that minimizes $Q(N)$. Therefore, it is obtained as a solution of the set of simultaneous equations

$$\partial Q(N)/\partial \kappa_p = 0 \qquad (p = 1, 2, \ldots, n). \tag{7.27}$$

Figure 7.4 shows the routine of the synthesis process just outlined.

7.3.4 Modification of κ_p by the Newton–Raphson Method

As shown at the bottom of Fig. 7.4, the Newton–Raphson method [7] is used in the iterative modification of κ_p. This method will first be described for a simple scalar equation.

Suppose the numerical solution of

$$G(\kappa) = 0 \tag{7.28}$$

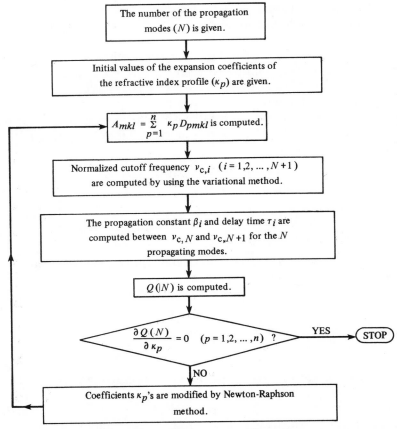

Fig. 7.4   Flow diagram of the synthesis of the optimum refractive-index profile [5].

is to be obtained. In the Newton–Raphson method, we first assume an initial value κ_0 and modify κ according to

$$\kappa_{h+1} = \kappa_h - [G(\kappa_h)/G'(\kappa_h)]d, \tag{7.29}$$

where h denotes the number of repetitions and d is a deceleration factor to prevent oscillations of the solution.

In the present case, the solution of the simultaneous equations

$$\mathbf{G(K)} = \begin{bmatrix} G_1(\mathbf{K}) \\ G_2(\mathbf{K}) \\ \vdots \\ G_n(\mathbf{K}) \end{bmatrix} = \mathbf{0} \tag{7.30}$$

must be obtained, where $\mathbf{K} = (\kappa_1, \kappa_2, \ldots, \kappa_n)^{\mathrm{T}}$ is an unknown vector quantity. In this case the modification formula corresponding to Eq. (7.29) is [8]

$$\mathbf{K}_{h+1} = \mathbf{K}_h - \mathbf{J}^{-1}(\mathbf{K}_h)\mathbf{G}(\mathbf{K}_h)d. \tag{7.31}$$

In this equation $\mathbf{J}^{-1}$ is the inverse matrix of $\mathbf{J}$, the Jacobian of $\mathbf{G}(\mathbf{K})$ with respect to $\mathbf{K}$ defined as

$$\mathbf{J}(\mathbf{K}) = \begin{bmatrix} J_{11}(\mathbf{K}) & \cdots & J_{1n}(\mathbf{K}) \\ \vdots & & \vdots \\ J_{n1}(\mathbf{K}) & \cdots & J_{nn}(\mathbf{K}) \end{bmatrix}, \tag{7.32}$$

where

$$J_{pq}(\mathbf{K}) = \partial G_p(\mathbf{K})/\partial \kappa_q \qquad (p, q = 1, 2, \ldots, n). \tag{7.33}$$

Under proper conditions, $\mathbf{K}_h$ converges to the solution of Eq. (7.30), which satisfies

$$\mathbf{J}^{-1}(\mathbf{K})\mathbf{G}(\mathbf{K}) = \mathbf{0}; \tag{7.34}$$

therefore,

$$\mathbf{K}_{h+1} - \mathbf{K}_h \rightarrow \mathbf{0}. \tag{7.35}$$

In the present problem, each line of the simultaneous equation Eq. (7.30) is given, from Eq. (7.27), as

$$G_p(\mathbf{K}) = \partial Q(N)/\partial \kappa_p \qquad (p = 1, 2, \ldots, n). \tag{7.36}$$

In the actual computation of $Q(N)$, the integrals in Eq. (5.2.31) are replaced by summations to save computer time; i.e., we choose M sampling points $v_j (j = 1, 2, \ldots, M)$ between $v_{c,N}$ and $v_{c,N+1}$ and use

$$Q(N) = \frac{1}{M} \sum_{j=1}^{M} \sigma(v_j) \tag{7.37}$$

as the approximate measure of the multimode dispersion instead of Eq. (5.231). Therefore, each element of $\mathbf{G}(\mathbf{K})$ and $\mathbf{J}(\mathbf{K})$ is, from Eqs. (7.33), (7.36), and (7.37),

$$G_p(\mathbf{K}) = \frac{1}{M} \sum_{j=1}^{M} \frac{\partial \sigma(v_j)}{\partial \kappa_p}, \tag{7.38}$$

$$J_{qp}(\mathbf{K}) = \frac{1}{M} \sum_{j=1}^{M} \frac{\partial^2 \sigma(v_j)}{\partial \kappa_q \partial \kappa_p} \qquad (p, q = 1, 2, \ldots, n). \tag{7.39}$$

The details of the computation of $G_p(\mathbf{K})$ and $J_{pq}(\mathbf{K})$ are described in Appendix II of Okamoto and Okoshi [5] and are omitted here because of space restrictions. Substituting those values into Eq. (7.31), we can compute $\mathbf{K}_{h+1}$ from $\mathbf{K}_h$. Repetition of such a process finally leads to the optimum $\mathbf{K}$, i.e., the optimum κ_p set.

7.3.5 Result of the Optimum-Profile Synthesis

To show the feasibility of the preceding process, an example of the synthesis will be described for the following parameters [5]:

$N = 10$ N is the number of propagating LP (linearly polarized) modes
          The total number of the conventional modes (having different polarization and/or field configuration) included in the 10 LP modes is 34. See Table 4.3.

$n = 5$ n is the number of terms representing the refractive-index profile. That $n = 5$ means that terms up to r^{10} are considered.

$M = 3$ M is the number of sampling points between $v_{c,10}$ and $v_{c,11}$.

$L = 10$ L is the number of terms representing the electric field in the core.

$y = 0.3$ y is the difference in material dispersions in the core and the cladding. See Eq. (4.100).

$n_1 = 1.5$ n_1 is the maximum refractive index in the core.

$\Delta = 0.01$ For Δ, see Eq. (7.23).

The starting profile is a "quadratic distribution with a valley having depth equal to half the peak," i.e., $\kappa_1 = 1.5$, $\kappa_2 = \kappa_3 = \kappa_4 = \kappa_5 = 0$. This profile is relatively good as indicated from our previous work on α-power refractive-index distributions in the core (see Fig. 5.11e).

The gradual change in the profile during the iterative synthesis process is shown in Fig. 7.5, where the parameter h denotes the number of iterations performed. The corresponding variations of the expansion coefficients are shown in Fig. 7.6 as functions of h. It is found that the coefficients for higher-order terms (x^4, x^6, x^8, x^{10}) drastically increase, but the resultant profiles are similar to the original. The principal difference is that the sharp valley is smoothed. (The computer time required was about 20 s for one cycle of the analysis, estimation, and modification, when an HITAC 8700/8800 was used [5].)

The corresponding variation of the normalized delay time between $v_{c,10}$ and $v_{c,11}$ is shown in Fig. 7.7. The variations of a maximum delay difference

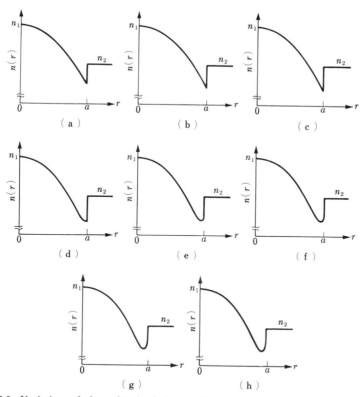

Fig. 7.5 Variation of the refractive-index profile during the synthesis process (after Okamoto and Okoshi [5]).

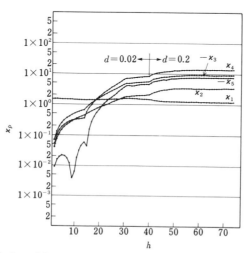

Fig. 7.6 Variation of the expansion coefficients during the synthesis process [5].

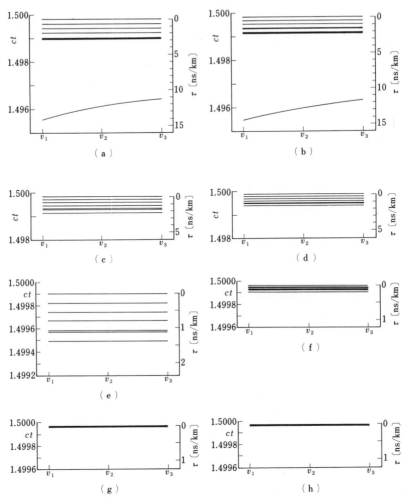

Fig. 7.7   Reduction of the normalized delay time [5]. Note that the ordinate is expanded in (e)–(h). The parameter h denotes the number of repetitions. The left and right edges of the graphs correspond to $v_{c,10}$ and $v_{c,11}$.

between modes T and the square root of the measure $Q(N)$ defined by (7.37) are shown in Fig. 7.8. In Fig. 7.8, one might notice a sudden decrease of the dispersion at $h = 14$. Such a phenomenon is sometimes encountered in the Newton–Raphson solution; it is due to an accidental, lucky coincidence. Note that the expansion coefficients do not show any drastic change at $h = 14$ (see Fig. 7.6).

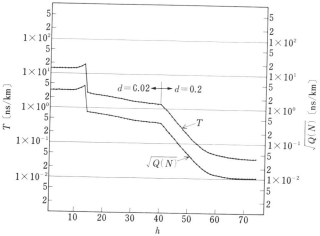

Fig. 7.8 Variation of the maximum intermodal delay difference T and the measure of the multimode dispersion Q during the synthesis process (after Okamoto and Okoshi [5]).

It is found that both T and $\sqrt{Q(N)}$ are reduced dramatically to about 3×10^{-3} times their starting values. Note that the starting values are much lower than the values obtained with a uniform-core fiber. The typical values of T for $\Delta = 0.01$ and $N = 10$ are approximately 50 ns/km for the uniform-core fiber, 14 ns/km for the starting profile,* and 40 ps/km for the really optimized profile. The last value is lower than the material dispersion obtained with an ordinary silica fiber and a typical multimode semiconductor laser with a wavelength spread of 1 nm; for such a combination, τ_m is typically 80 ps/km at $\lambda = 850$ mm (see Table 4.5).

7.3.6 Validity of Result

By the curve-fitting method, it is found that the optimum refractive-index profile can be approximated as

$$n^2(r) = n_1^2[1 - 4.00\,\Delta(r/a)^{2.27}] \tag{7.40}$$

for $0 \le r/a \le 0.9$, when $y = 0.3$. On the other hand, when $y = 0$, the approximate profile is

$$n^2(r) = n_1^2[1 - 4.04\,\Delta(r/a)^{1.97}] \tag{7.41}$$

---

* This value may seem large; the reason is that the highest order mode is included in the calculation of T. If we exclude the highest order mode, T is 3 ns/km for the starting profile.

for $0 \leq r/a \leq 0.9$. These approximate formulas are in good agreement with the result of Olshansky and Keck's analysis [9], and also with the optimum parameters obtained in Sections 5.4.2 and 5.7.1, i.e., $\alpha = 2 + y$ and $\rho = 2$. Moreover, we should note that in Fig. 5.12b where the effect of "filling up the valley" is investigated, the least dispersion is obtained at $\rho_0 = 1.5$ when $N = 10$. The curve for $\rho_0 = 1.5$ in Fig. 5.12b shows a very close resemblance to Fig. 7.5h; the only difference is that the profile is smoothed in the latter.

A possible problem inherent to such a iterative synthesis is the presence of many optima, i.e., the possibility of arriving at different "optimum" profiles when the starting profiles are different. Mathematically, such a possibility cannot be denied in the present case. However, another iterative synthesis starting from a different profile ($\kappa_1 = 0.1$, $\kappa_2 = \kappa_3 = \kappa_4 = \kappa_5 = 0$) finally led to an optimum profile identical to Fig. 7.5h. It is believed that the same profile is obtained unless we start from an intentionally peculiar (e.g. sawtoothlike) profile.

An example of the synthesis has been described for an optical fiber in which 10 modes propagate. Of course, the same methods may be used for any number of modes N, and probably similar result will be obtained. However, the computer time required is proportional to N.

Finally, we should note that the dispersion is reduced by 10^{-1} between $h = 50$ and $h = 70$, whereas the profiles show very little difference between these cases. In the present state-of-the-art, such a little difference of the profile can never be controlled. Establishement of a very fine measurement and control techniques of the profile is desired [8].

References

1. S. Kawakami and S. Nishida, Characteristics of a doubly clad optical fiber with a low-index inner cladding, *IEEE J. Quantum Electron.* **QE-10**, No. 12, 879–887 (1974).
2. S. Kawakami and S. Nishida, Purterbation theory of a doubly clad optical fiber with a low-index inner cladding, *IEEE J. Quantum Electron.* **QE-11**, No. 4, 130–138 (1975).
3. D. Gloge and E. A. J. Marcatili, Multimode theory of graded-core fibers, *Bell Syst. Tech. J.* **52**, No. 9, 1563–1578 (1973).
4. K. Okamoto and T. Okoshi, Analysis of wave propagation in optical fibers having core with α-power refractive index distribution and uniform cladding, *IEEE Trans. Microwave Theory Tech.* **MTT-24**, No. 7, 416–421 (1976).
5. K. Okamoto and T. Okoshi, Computer-aided synethesis of the optimum refractive index profile for a multimode fiber, *IEEE Trans. Microwave Theory Tech.* **MTT-25**, No. 3, 213–221 (1977).
6. K. Furuya, Y. Suematsu, T. Tanaka, and S. Ishikawa, An ideal refractive-index distribution and mode filter for band broadening of multimode optical fibers, *IEEE MTT-S Symp. Cherry Hill*, Pennsylvania, Paper No. 2–5, June 14–16 (1976).
7. E. Polak, "Computational Methods in Optimization." Academic Press, New York, 1971.

8. A. Ben-Israel, A Newton–Raphson method for the solution of systems of equations, *J. Math. Anal. Appl.* **15**, 243–252 (1966).

9. R. Olshansky and D. B. Keck, Pulse broadening in graded-index optical fibers, *Appl. Opt.* **15**, No. 2, 483–491 (1976).

10. T. Okoshi, Optimum profile design of optical fibers and related requirements for profile measurement and control (invited), IOOC'77, Technical Digest No. C2-1, July. 18–20, Tokyo (1977).

8 | Optical Fibers Having Structural Fluctuations

In an optical fiber in which structural fluctuations (e.g., random variation of the diameter or undulation of the axis) are present along its length, mode conversion takes place. As a result, multimode dispersion is reduced. However, transmission loss increases due to the presence of the axial fluctuation. Thus, there exists a limitation of dispersion reduction which uses the mode-conversion effect.

Propagation characteristics of multimode optical fibers having structural fluctuations along the axis are described; a uniform-core fiber is considered as the simplest example.

8.1 Introduction

In the preceding chapters, propagation characteristics of optical fibers have been studied and their design principles presented. However, we have considered only axially uniform fibers.

In actual fibers, structural fluctuations are present in the axial direction. For example, both the fiber diameter and refractive-index profile fluctuate along the fiber length. The fiber axis undulates more or less along its length. In such axially fluctuating fibers, light scattering takes place, thereby increasing transmission loss. Thus the axial fluctuation is harmful on one hand; but advantageous on the other because of mode conversion generation, exchanging light energy from one mode to another. As a result, a wave packet experiences a number of different modes while traveling from the launch end to the receiving end, and multimode dispersion is reduced due to the effect of averaging various group velocities of those different modes.

Experimental data showing the reduction of multimode dispersion due to mode conversion is presented first. Next, mode conversion and its effect on the impulse response are calculated for a uniform-core fiber to substatiate the experimental data.

8.2.   Mode Conversion due to Axial Fluctuation

8.2.1   Improvement of Dispersion Characteristics by Mode Conversion

As shown in Chapters 3–5, the impulse response width of a multimode fiber is proportional to the propagation distance z. However, it is often observed in actual fibers that beyond a certain propagation distance, the impulse response width becomes proportional to $\sqrt{z}$. Figure 8.1 shows a typical example of the relation between the distance and the impulse response width reported by Chinnock et al. [1]; the $\sqrt{z}$ dependence is observed for $z > 500$ m.

The $\sqrt{z}$ dependence stems from mode conversion; a qualitative explanation follows. We consider that a transmitted optical pulse is an ensemble of many energy packets. At the launch end, those packets are distributed across the propagating modes and start traveling with their individual group velocities. If the fiber structure is ideal and no axial fluctuations are present, the group velocity of each packet will be constant throughout its travel, and the multimode dispersion as computed in preceding chapters will be

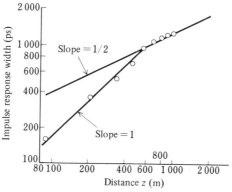

Fig. 8.1   Relation between the propagation distance and the impulse response width (after Chinnock et al. [1]).

observed. The impulse response width will be proportional to z. However, when axial fluctuations are present, mode conversion takes place, and each energy packet sees various modes, and hence, experiences various group velocities. Thus, delay time variances for the energy packets become smaller. It was predicted by Personik as early as 1971 that the impulse response width would become proportional to $\sqrt{z}$ beyond a certain propagation distance. This prediction resulted from his consideration of the two-mode model [2].

As described in the Introduction, axial fluctuation inevitably increases transmission loss. Therefore, to take advantage of mode conversion effects, a compromise must be made between impulse (or frequency) response and transmission loss. A proper compromise depends on the state of the art, and also on the specifications of the optical fiber communication system to be constructed. When the designer feels that the maximum separation between optical communication repeaters is not limited by the transmission loss but by the bandwidth, he may give axial fluctuation to the fiber and take advantage of the band-broadening effect, trading off transmission loss. When he feels that the repeater separation is limited by transmission loss, he should of course try to reduce axial fluctuation in the fiber.

At present, many researchers are concerned with the band-broadening effect due to mode conversion because the repeater separation is often limited by the bandwidth requirement.

8.2.2   Coupled Power Equations

Mode coupling phenomena are usually analyzed with the coupled mode equations,

$$\partial a_v / \partial z = \sum_{\mu=1}^{N} C_{v\mu} a_\mu, \tag{8.1}$$

where a_v ($v = 1, 2, \ldots$) and $C_{v\mu}$ denote the amplitude of the vth mode and the conversion coefficient from the μth to the vth mode, respectively. Coefficients $C_{v\mu}$ satisfy

$$C_{v\mu}^* = -C_{\mu v}, \tag{8.2}$$

$$C_{vv} = -j\beta_v, \tag{8.3}$$

where the asterisk (*) denotes a complex conjugate, and β_v is the propagation constant of the vth mode.

Equation (8.1) can be derived directly from Maxwell's equations [3], and its physical implication can be easily understood. The coefficients $C_{v\mu}$ are called amplitude coupling coefficients. These can be computed if the geometrical parameters of the fiber are given exactly; typically,

$$C_{v\mu} = K_{v\mu} f(z), \tag{8.4}$$

where $f(z)$ corresponds to the axial fluctuation; sometimes it is equal to the shape of the geometrical fluctuation.

In principle, all the mode coupling phenomena in an optical fiber could be calculated by using Eq. (8.1). However, in actual fiber analysis, this equation cannot be used in its present form because $f(z)$ is given only in a statistical manner, and hence the phase differences between modes cannot be determined along the fiber length.

To avoid this difficulty, Marcuse proposed the use of a coupling equation in terms of the ensemble average of the power of each mode [4]

$$dP_v/dz = -\alpha_v P_v + \sum_{\mu=1}^{N} d_{v\mu}(P_\mu - P_v), \tag{8.5}$$

where α_v is the attenuation constant due to absorption and scattering, and $d_{v\mu}$ is a quantity called the power coupling coefficient defined as

$$d_{v\mu} = |K_{v\mu}|^2 \int_{-\infty}^{\infty} \langle f(z)f(z-u)\rangle e^{j\Delta\beta_{v\mu}u}\,du. \tag{8.6}$$

In this definition,

$$\langle f(z)f(z-u)\rangle = \lim_{L\to\infty}(1/2L)\int_{-L}^{L} f(z)f(z-u)\,dz \tag{8.7}$$

is the autocorrelation function of $f(z)$, and $\Delta\beta_{v\mu}$ is the difference between propagation constants of two modes,

$$\Delta\beta_{v\mu} = \beta_v - \beta_\mu. \tag{8.8}$$

We should note that the "weakly coupling assumption" is made in deriving Eq. (8.5).

Equations (8.5) are called "coupled power equations." A significant fact of these equations is that the coupling coefficient $d_{v\mu}$ is expressed, not as functions of z, but as functions of the difference of propagation constants $\Delta\beta_{v\mu}$. Furthermore, according to the Wiener–Khintchine theorem,

$$\int_{-\infty}^{\infty} \langle f(z)f(z-u)\rangle e^{j\Delta\beta_{\mu v}u}\,du = \lim_{L\to\infty}(1/2L)\left|\int_{-L}^{L} f(z)e^{j\Delta\beta_{v\mu}z}\,dz\right|^2. \tag{8.9}$$

Hence, the power coupling coefficient is proportional to the power spectral density of $f(z)$ at angular spatial frequency $\Delta\beta v\mu$.

We consider here a coupling between two adjacent [mth and $(m + 1)$th] modes, whose β's are close to each other. The coupling coefficient is proportional to the angular spatial frequency component equal to $\Delta\beta_{m, m+1} = \beta_m - \beta_{m+1}$ of the power spectral density of the structural fluctuation along the fiber length. In other words, the component of the fluctuation having a period $(2\pi/\Delta\beta_{m, m\pm1})$ contributes to the coupling. As shown in Section 6.3.5 [see Eq. (6.37)], this period is that of the ray undulation corresponding to the mth mode. It is also equal to the distance over which the phase difference between the mth and $(m + 1)$th modes varies by 2π.

8.2.3   Degeneration among Modes

In the coupled power equations just described, we denote a mode by a single index. Of course, actual fiber modes are characterized by two indices: one for the radial mode and the other for the azimuthal (rotational) mode. Here it is understood that the single mode corresponds to a pair.

From a practical standpoint, we can actually denote different modes by one integer value. We should note first that in the present analysis we are not concerned with the field distribution of propagating modes but only with their propagation velocities. If two modes have identical β's (and hence the same $d\beta/d\omega$), they need not be considered separately. As an example, we first consider a quadratic-core fiber ($\alpha = 2$). In this case, as shown in Appendix 8A.1, modes having identical $(2l + m)$(l and m denote the radial and rotational mode numbers values are degenerate [5].

Next, we consider a uniform-core fiber. When the refractive-index difference between core and cladding is small, the propagation constant β is approximately, from Eq. (4.78) [6],

$$\beta \simeq kn_2\{1 + \Delta - \Delta(u^2/v^2)\},\tag{8.10}$$

where k is the wave number in vacuum and Δ is the relative index difference defined in Section 3.2.2, in terms of the core index n_1 and cladding index n_2,

$$\Delta \equiv (n_1^2 - n_2^2)/2n_1^2 \simeq (n_1 - n_2)/n_1.\tag{8.11}$$

On the other hand, at frequencies far from cutoff, the dispersion equation is

$$J_m(u) \simeq 0,\tag{8.12}$$

because w becomes very large, and hence, the right-hand side of Eq. (4.75) becomes very large. (Note that K_{m-1}/K_m approaches unity for large w. See Fig. 4.2b.) Thus, if we denote the normalized transverse wave number for the (m, l) mode by u_{ml}, then u_{ml} is insensitive to the normalized frequency v to a first-order approximation.

Equation (8.10) shows that modes having identical u_{ml} values are degenerate. On the other hand, an approximate solution of Eq. (8.12) is

$$u_{ml} \simeq (\pi/2)(2l + m), \tag{8.13}$$

indicating that modes having identical $(2l + m)$ values are also degenerate in uniform-core fibers [7].

The quadratic-core and uniform-core profiles are typical in practical fibers but are quite different. Therefore, henceforth, we shall consider that for any type of the profile, the mode-group number

$$m' = 2l + m \tag{8.14}$$

can be used in dealing with the intermodal coupling phenomena.

The number of combinations of l and m having identical m' value is the integer smaller than but closest to $m'/2$. Therefore, considering the degeneration between $\text{HE}_{(m+1)l}$ and $\text{EH}_{(m-1)l}$ modes and the presence of horizontal and vertical polarizations (see Table 4.3), we may conclude that approximately $2m'$ conventional modes are included in the m'th mode group as just defined.

8.2.4   Conversion between Adjacent Modes

Typically, low-order modes radiate from a fiber end at a small angle relative to the fiber axis. On the other hand, the high-order modes radiate at large angles. To investigate the nature of the intermodal coupling, Gloge [7] measured the spread of the exit radiation pattern as a function of fiber length. At the launch end only lower-order modes were excited. He found that the power moved gradually from the lower to the higher modes along the fiber length; this fact suggests that coupling takes place principally between "close" modes.

To simplify the problem, we shall assume, following Gloge [7, 8], that coupling takes place only between adjacent modes, and rewrite the coupled power equation (8.5) as

$$m \frac{dP_m}{dz} = -m\alpha_m P_m + md_m(P_{m+1} - P_m) + (m-1)d_{m-1}(P_{m-1} - P_m). \tag{8.15}$$

In this equation, the mode-group number m' in Eq. (8.14) is rewritten as m (prime removed) because it appears rather frequently in the subsequent equations. The symbol P_m denotes the average power of the mth mode group, and d_m and d_{m-1} are the coupling coefficients between the $(m+1)$th and mth mode groups and between the mth and $(m-1)$th mode groups, respectively. The multiplication of each term in Eq. (8.15) by m or $(m-1)$

corresponds to the fact that $2m$ conventional modes are included in the mth mode group. Hereafter we call the mth mode group the "mth mode".

We introduce here a new normalized parameter proportional to m,

$$\theta = m\lambda/4an_1 \simeq u/an_1k, \qquad (8.16)$$

where a denotes the core radius, λ the wavelength in vacuum, and $k = 2\pi/\lambda$. It can be shown that the new parameter θ thus defined gives approximately the angle made by the central axis of the fiber and the ray corresponding to the mth mode crossing the axis [7]. Thus, it is approximately equal to $(1/n_1)$ times (where n_1 denotes the core index) the exit radiation angle.

When a very large number of modes are propagating, the mode number m can be regarded as a continuous quantity. In such a case, θ becomes also continuous, and a partial derivative

$$(P_{m+1} - P_m)/(\theta_{m+1} - \theta_m) = \partial P_m/\partial\theta \qquad (8.17)$$

can be defined approximately. Therefore, Eq. (8.15) becomes

$$m\frac{dP_m}{dz} = -m\alpha_m P_m + \Delta\theta\left\{md_m\frac{\partial P_m}{\partial\theta} - (m-1)d_{m-1}\frac{\partial P_{m-1}}{\partial\theta}\right\}, \qquad (8.18)$$

where

$$\Delta\theta = \theta_{m+1} - \theta_m. \qquad (8.19)$$

Equation (8.18) can be rewritten as

$$\frac{dP_m}{dz} = -\alpha_m P_m + (\Delta\theta)^2\frac{1}{m}\frac{\partial}{\partial\theta}\left(md_m\frac{\partial P_m}{\partial\theta}\right), \qquad (8.20)$$

or by using θ instead of m, as

$$\frac{dP}{dz} = -\alpha(\theta)P + (\Delta\theta)^2\frac{1}{\theta}\frac{\partial}{\partial\theta}\left\{\theta d(\theta)\frac{\partial P}{\partial\theta}\right\}. \qquad (8.21)$$

To proceed with the analysis, we have to assume the θ-dependence of the attenuation constant $\alpha(\theta)$ and that of the power coupling coefficient $d(\theta)$. We assume that $\alpha(\theta)$ can be expressed in the form

$$\alpha(\theta) = \alpha_0 + A\theta^2. \qquad (8.22)$$

The first term α_0 gives an attenuation independent from the mode number; absorption and scattering in the core are included here. The second term gives the attenuation due to scattering at the core–cladding boundary; the form $A\theta^2$ is based on the theoretical analysis result [6] that the power density at the core–cladding boundary is approximately proportional to m^2. In the subsequent analysis, the first term will be neglected because only the second term is essential when considering the attenuation directly relevant to mode conversion. We assume that the power coupling coefficient is given as a constant regardless of the mode number, i.e.,

$$d(\theta) = d_0. \tag{8.23}$$

If we substitute the foregoing relations into Eq. (8.21) and use a new parameter

$$D = (\Delta\theta)^2 d_0 = (\lambda/4an_1)^2 d_0, \tag{8.24}$$

then Eq. (8.21) is rewritten as

$$\frac{dP}{dz} = -A\theta^2 P + \frac{1}{\theta}\frac{\partial}{\partial\theta}\left(\theta D \frac{\partial P}{\partial\theta}\right). \tag{8.25}$$

On the other hand, the chain rule gives

$$\frac{dP}{dz} = \frac{\partial P}{\partial z} + \frac{\partial P}{\partial t}\frac{dt}{dz}, \tag{8.26}$$

where dt/dz is the inverse group velocity of the mode. As shown in Appendix 8A.2, it is expressed as

$$dt/dz \simeq n_1(1 + \theta^2/2)/c, \tag{8.27}$$

where c is the light velocity in vacuum. Thus, Eq. (8.25) becomes

$$\frac{\partial P}{\partial z} = -A\theta^2 P - \frac{n_1}{2c}\theta^2\frac{\partial P}{\partial t} + \frac{1}{\theta}\frac{\partial}{\partial\theta}\left(\theta D \frac{\partial P}{\partial\theta}\right). \tag{8.28}$$

Note that the delay n_1/c common to all modes is deducted from Eq. (8.27).

Equation (8.28) is the basic differential equation giving the relation among $\partial P/\partial z$, $\partial P/\partial t$, and $\partial P/\partial\theta$.

8.2.5 Impulse Response

The impulse response can be obtained by solving Eq. (8.28) with the proper initial conditions [8]. We first define the Laplace transform of P,

$$p(\theta, z, s) = \int_0^\infty P(\theta, z, t)e^{-st}\, dt \tag{8.29}$$

and rewrite Eq. (8.28) as

$$\frac{\partial p}{\partial z} = -A\sigma^2\theta^2 p + \frac{1}{\theta}\frac{\partial}{\partial\theta}\left(\theta D \frac{\partial p}{\partial\theta}\right), \tag{8.30}$$

where

$$\sigma = (1 + n_1 s/2cA)^{1/2}. \tag{8.31}$$

Note that σ includes the variable s.

Next, we assume that at the launch end the optical power is distributed to the propagating modes in a Gaussian distribution with respect to θ, so that

$$p_{in}(\theta, s) = f(0, s) \exp(-\theta^2/\Theta_0^2). \tag{8.32}$$

Under this assumption, the solution of Eq. (8.30) can also be assumed to be Gaussian; therefore, we may write

$$p(\theta, z, s) = f(z, s) \exp\{-\theta^2/\Theta^2(z, s)\}. \tag{8.33}$$

In this equation, $f(z, s)$ and $\Theta(z, s)$ express the variation of the pulse waveform and the energy distribution spread among modes, respectively.

Substituting Eq. (8.33) into (8.30) and equating each power of θ to zero, we obtain

$$d\Theta/dz = (-A\sigma^2/2)\Theta^3 + (2D/\Theta), \tag{8.34}$$

$$df/dz = -4Df/\Theta^2. \tag{8.35}$$

The solution of Eq. (8.34) is given as

$$\Theta^2 = \begin{cases} \Theta_\infty^2[\tanh\sigma\gamma_\infty(z + z_0)]/\sigma & \text{(when} \quad \Theta_\infty \geq \Theta_0), \\ \Theta_\infty^2[\coth\sigma\gamma_\infty(z + z_0)]/\sigma & \text{(when} \quad \Theta_\infty < \Theta_0), \end{cases} \tag{8.36}$$

where

$$\Theta_\infty = (4D/A)^{1/4}, \tag{8.37}$$

$$\gamma_\infty = (4DA)^{1/2}, \tag{8.38}$$

and z_0 is a constant determined by the launch condition. [The physical implications of Θ_∞ and γ_∞ will be stated following Eq. (8.41).] Substituting Eq. (8.36) into (8.35), we obtain

$$f = \begin{cases} f_0/\sinh\sigma\gamma_\infty(z + z_0) & (\Theta_\infty \geq \Theta_0), \\ f_0/\cosh\sigma\gamma_\infty(z + z_0) & (\Theta_\infty < \Theta_0), \end{cases} \tag{8.39}$$

where f_0 is again a constant determined by the launch condition.

If we assume that $\Theta_\infty \geq \Theta_0$, and determine z_0 and f_0 by using Eq. (8.32), we obtain, from Eqs. (8.36) and (8.39),

$$\Theta^2(z, s) = \frac{\Theta_\infty^2}{\sigma} \frac{\sigma\Theta_0^2 + \Theta_\infty^2 \tanh\sigma\gamma_\infty z}{\Theta_\infty^2 + \sigma\Theta_0^2 \tanh\sigma\gamma_\infty z}, \tag{8.40}$$

$$f(z, s) = \frac{f(0, s)\sigma\Theta_0^2}{\Theta_\infty^2 \sinh\sigma\gamma_\infty z + \sigma\Theta_0^2 \cosh\sigma\gamma_\infty z}. \tag{8.41}$$

When continuous light is launched, $s = 0$ and hence $\sigma = 1$. In this case Eq. (8.40) suggests that the steady-state modal spread $\Theta(z, 0)$ varies from Θ_0 to Θ_∞ monotonously as z increases. Thus, Θ_∞ gives the equilibrium modal spread. The parameter γ_∞ gives the variation rate of the modal spread toward that equilibrium value [see Eq. (8.40)]. However, substituting $\sigma = 1$ and $z \to \infty$ into Eq. (8.41), we find that γ_∞ is the attenuation constant at the equilibrium state. Since it is assumed that $\alpha_0 = 0$ in Eq. (8.22) and hence the attenuation can be attributed only to the structural fluctuation at the core–cladding boundary, γ_∞ denotes the loss accompanying the energy conversion among modes. The nature of Θ_∞ and γ_∞ is discussed in more detail in Appendix 8A.3.

In the subsequent discussions, we consider only those cases in which $\Theta_0 = \Theta_\infty$, i.e., the input light power is distributed to the propagating modes in the "equilibrium ratio." We first consider a very short fiber. The Laplace transform of the total power is

$$q(z, s) = 2\pi \int_0^\infty p(\theta, z, s)\theta \, d\theta. \tag{8.42}$$

Since we can approximate as $\sinh \sigma\gamma_\infty z \simeq \tanh \sigma\gamma_\infty z \simeq \sigma\gamma_\infty z$ and $\cosh \sigma\gamma_\infty z \simeq 1$, we obtain, from Eqs. (8.33) and (8.40)–(8.42),

$$q(z, s) = \frac{\pi f(0, s)\Theta_0^2}{(1 + \gamma_\infty z)(1 + n_1 \Theta_0^2 zs/2c)}. \tag{8.43}$$

Substituting $f(0, s) = 1$ into the preceding equation and computing its inverse Laplace transform, we obtain the impulse response as

$$Q(z, t) = \frac{2c\pi}{n_1 z(1 + \gamma_\infty z)} \exp(-2ct/n_1 \Theta_0^2 z). \tag{8.44}$$

Equation (8.44) shows that the impulse response of a short fiber is an attenuating Gaussian function whose spread is proportional to the fiber length z; i.e., the effect of the mode conversion does not appear in the impulse response of a very short fiber; this conclusion is quite natural.

Next, we consider a very long fiber. Since we can approximate in this case that $\tanh \sigma\gamma_\infty z \simeq 1$ and $\sinh \sigma\gamma_\infty z \simeq \cosh \sigma\gamma_\infty z \simeq \frac{1}{2}\exp \sigma\gamma_\infty z$, we obtain

$$q(z, s) = [2\pi\Theta_0^2/(1 + \sigma)]\exp(-\sigma\gamma_\infty z). \tag{8.45}$$

Assuming that $\gamma_\infty z > 1$ and computing the inverse Laplace transform of Eq. (8.45), we obtain the impulse response as [8]

$$Q(z, t) = \Theta_0^2 \sqrt{\frac{\pi}{Tt}} \left(\frac{t}{\gamma_\infty zT} + \frac{1}{2}\right)^{-1} \exp\left(-\frac{\gamma_\infty^2 z^2 T}{4t} - \frac{t}{T}\right), \tag{8.46}$$

where

$$T = n_1/2cA = (n_1/2c)(\Theta_0^2/\gamma_\infty). \tag{8.47}$$

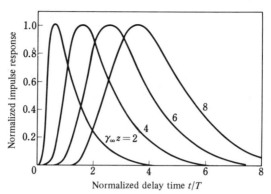

Fig. 8.2   Impulse response $Q(z, t)$ for various values of $\gamma_\infty z$ (after Gloge [8], Copyright 1973, American Telephone and Telegraph Company; reprinted by permission).

Figure 8.2 shows the impulse response $Q(z, t)$ with $\gamma_\infty z$ as a parameter. The ordinate is normalized so that the peak value becomes unity. It is found that as $\gamma_\infty z$ increases, the impulse response approaches a symmetrical Gaussian function.

8.2.6   Delay Time and Pulse Broadening

So far we have investigated the shape of the impulse response. We are now ready to compute the z dependence of the pulse spread as shown in Fig. 8.1.

The relation between the moment of a function and its Laplace transform is used in the following calculation. Generally, the mth-order moment of a probability density function $Q(t)$ around its average value δ can be expressed as

$$\frac{\int_0^\infty (t - \delta)^m Q\, dt}{\int_0^\infty Q\, dt} = (-1)^m \left. \frac{\partial^m \ln q}{\partial s^m} \right|_{s=0}, \tag{8.48}$$

where $q(s)$ is the Laplace transform of $Q(t)$ [8]. The average δ is

$$\delta = \frac{\int_0^\infty tQ\, dt}{\int_0^\infty Q\, dt} = -\left. \frac{\partial \ln q}{\partial s} \right|_{s=0}. \tag{8.49}$$

The variance of $Q(t)$ is obtained by putting $m = 2$ into Eq. (8.48).

On the other hand, the Laplace transform of the impulse response of an optical fiber is, from Eqs. (8.33) and (8.40)–(8.42),

$$q(z, s) = \pi \Theta_0^2 (\sigma \sinh \sigma\gamma_\infty z + \cosh \sigma\gamma_\infty z)^{-1}. \tag{8.50}$$

Substituting this into Eq. (8.49), we obtain the average delay of the impulse response,

$$\delta_Q = (T/2)\{\gamma_\infty z + \tfrac{1}{2}(1 - e^{-2\gamma_\infty z})\}, \tag{8.51}$$

where T is a constant defined by Eq. (8.47). [To obtain the overall delay time from launch end to receiving end, we have to add $(n_1/c)z$ to Eq. (8.51).] The impulse response variance can be obtained by putting $m = 2$ and Eq. (8.50) into (8.48). Since this variance is considered to be equal to the square of the pulse width τ_Q, we obtain

$$\tau_Q = (T/2)\{\gamma_\infty z(1 - 2e^{-2\gamma_\infty z}) + \tfrac{3}{4} - e^{-2\gamma_\infty z} + \tfrac{1}{4}e^{-4\gamma_\infty z}\}^{1/2}. \tag{8.52}$$

The solid curve in Fig. 8.3 shows the relation between the normalized propagation distance $\gamma_\infty z$ and the normalized width of the impulse response τ_Q/T computed using Eq. (8.52). In the region $z \ll 1/\gamma_\infty$, τ_Q/T approaches $\gamma_\infty z$ and hence is proportional to z; this fact suggests that the effect of the mode conversion is absent. On the contrary, the relation asymtotically approaches

$$\tau_Q = (T/2)(\gamma_\infty z)^{1/2} \tag{8.53}$$

as z exceeds $z_c = 1/4\gamma_\infty$ and increases further. Thus, the $\sqrt{z}$ dependence of the pulse spread shown in Fig. 8.1 has been proved.

If we are concerned only with the improvement of the impulse response of a multimode fiber, we should make the critical distance z_c (where $\sqrt{z}$ dependence starts) as short as possible. However, to make z_c shorter, the loss must inevitably increase because $z_c = 1/4\gamma_\infty$. Thus, we arrive at the

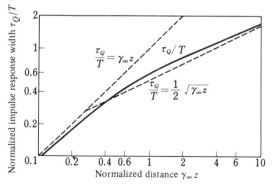

Fig. 8.3   Relation between normalized propagation distance $\gamma_\infty z$ and normalized impulse response width τ_Q/T (after Gloge [8], Copyright 1973, American Telephone and Telegraph Company; reprinted by permission).

important conclusion that improvement of the impulse response using mode conversion can be achieved only with the sacrifice of increased transmission loss.

8.3   Summary

Propagation characteristics of a multimode optical fiber having structural fluctuations along its length has been discussed. It has been shown that the impulse response can be improved by power conversion among modes, but with the sacrifice of increased transmission loss. Uniform-core fibers have mainly been considered; discussions on the quadratic-core fibers and α-power graded core fibers will be found in the literature [9, 10].

References

1.   E. L. Chinnock, L. G. Cohen, W. S. Solden, R. D. Standley, and D. B. Keck, The length dependence of pulse spreading in the CGW-Bell-10 optical fiber, *Proc. IEEE* **61**, No. 10, 1499–1500 (1973).

2.   S. D. Personick, Time dispersion in dielectric waveguides, *Bell Syst. Tech. J.* **50**, No. 3, 843–859 (1971).

3.   D. Marcuse, "Theory of Dielectric Optical Waveguides," p. 95. Academic Press, New York, 1974.

4.   D. Marcuse, Derivation of coupled power equations, *Bell Syst. Tech. J.* **51**, No. 1, 229–237 (1972).

5.   S. Kawakami and J. Nishizawa, An optical waveguide with the optimum distribution of the refraction index with reference to waveform distortion, *IEEE Trans. Microwave Theory Tech.* **MTT-16**, No. 10, 814–818 (1968)

6.   D. Gloge, Weakly guiding fibers, *Appl. Opt.* **10**, No. 10, 2252–2258 (1971).

7.   D. Gloge, Optical power flow in multimode fibers, *Bell Syst. Tech. J.* **51**, No. 8, 1767–1783 (1972).

8.   D. Gloge, Impulse response of clad optical multimode fibers, *Bell Syst. Tech. J.* **52**, No. 6, 801–816 (1973).

9.   D. Marcuse, Losses and impulse response of a parabolic index fiber with random bends, *Bell Syst. Tech. J.* **52**, No. 8, 1423–1437 (1973).

10.   R. Olshansky, Mode coupling effects in graded-index optical fibers, *Appl. Opt.* **14**, No. 4 935–945 (1975).

9 | Measurement of Refractive-Index Profile of Optical Fibers

Discussions in Chapters 3–7 revealed that group-delay characteristics of an optical fiber are determined principally by its refractive-index profile. Therefore, the development of precise and efficient measurement techniques of the refractive-index profile is an important technical task.

Various methods will be presented in as systematic a manner as possible. All of these methods have their own merits and demerits; in actual measurement, a proper choice must be made in accordance with the situation and requirement.

9.1 Introduction

In Chapters 3–7, the analysis and design of optical fibers have been studied. Our greatest concern has been the relation between the refractive-index profile and the group-delay characteristics; it has actually been revealed that the former is the principal factor determining the latter. Therefore, in the manufacture and evaluation of optical fibers, the measurement of the index profile is one of the most important steps; hence, the development of precise and efficient methods for its measurement is an important technical task.

Various methods developed and reported prior to 1976, which are tabulated in Table 9.1, will be described in some detail and compared. The ideal measuring method should satisfy the following five conditions; i.e., it should be

(1) nondestructive,
(2) applicable to any profile,
(3) of high accuracy,
(4) of high resolution, and
(5) easy in measurement and processing of data.

Table 9.1 Various Methods for Measuring Refractive-Index Profile of Optical Fibers and Performs, with Primary Classification According to Measurement Objective (Reports prior to 1976)

Measurement objective	Destructive or nondestructive		Method	Quantity obtained [accuracy]	Reference	Text section
Unclad fibers	ND	A	Scattering pattern (backward)	n [0.2%], r [5%]	Presby [1]	9.2.8
Uniform-core fibers	ND	B	Scattering pattern (forward)	n_{core} [± 0.003], r_{core}, r_{clad}	Watkins [2]	9.2.8
		C	Matching oil	n_{core}, n_{clad}	Liu [3]	9.5.4
Quadratic-core fibers	D	D	Far-field pattern	fourth-order-term coefficient	Kitano et al. [4]	9.5.3
		E	Near-field pattern	fourth-order-term coefficient	Suematsu et al. [5]	9.5.3
		F	Interference between reflections at both ends of fiber	fourth- and sixth-order-term coefficients	Rawson et al. [6]	9.3.6
		G	Step in near-field pattern	fourth-order-term coefficient	Maeda et al. [7]	9.5.3
Fibers having arbitrary profile	D	H	Microscopic interferometry using sliced samples	n at any point	Martin [8], Cherin et al. [9], Wonsievicz et al. [10]	9.3.2
		I	Reflection method	n at any point [$\sim 5\%$]	Ikeda et al. [11], Ueno et al. [12], Eickoff et al. [13] Shiraishi et al. [14], Aoki et al. [15].	9.4
		J	Near-field pattern	n at any point	Gloge et al. [16], Payne et al. [17], [18], Tanaka et al. [19]	9.5.1
		K	X-ray microanalysis (XMA)	n at any point	Osanai et al. [20]	9.5.2
	ND	L	Microscopic interferometry using as-drawn samples	n at any point (axial symmetry assumed)	Shiraishi et al. [21], Hunter et al. [22], Iga et al. [23]	9.3.3, 9.3.4
		M	Scattering pattern (forward)	n at any point (axial symmetry assumed)	Okoshi and Hotate [24–27]	9.2.1–9.2.7
Arbitrary profiles (qualitative)	D	N	Use of scanning electron microscope	Qualitative measurement	Burrus et al. [28]	9.5.5
Step-index preforms	ND	O	Scattering pattern	Two radii and two indices	Chigira et al. [29]	9.2.8
Arbitrary-profile preforms	ND	P	Interferometry with fringe-localization scheme	n at any point (axial symmetry assumed)	Okoshi and Hotate [30]	9.3.5

However, from Table 9.1 we see that none of these methods statisfies all these conditions; only methods L and M satisfy conditions (1)–(4) simultaneously, but they do not satisfy condition (5). In actual measurement, therefore, a proper choice must be made in accordance with the situation and requirement.

The descriptions in the following sections are arranged according to the kind of primary data collected; methods using a scattering pattern, those using an interference-fringe pattern, and those using a reflection coefficient will be described first; miscellaneous methods using other data will then be presented. This order of description is different from that of Table 9.1 in which the classification is according to the measurement objective (see the column "Text section").

9.2 Scattering-Pattern Method

The method in which the index profile is computed from the scattering pattern for an normally incident light beam is called the scattering-pattern method. This method satisfies conditions (1) and (2) mentioned in Section 9.1. Condition (3) is satisfied when the core is relatively thin. The resolution of condition (4) is the highest among all methods available; the spatial resolution limit is equal to the quarter wavelength (in the material) of the light used in the measurement; actual resolution is typically 0.2 μm. Hence, at present this is the only method available for measuring the profile of single-mode fibers with practical resolution.

9.2.1 Optical Setup for Measuring the Scattering Pattern

Figure 9.1 shows the basic part of the optical setup for measuring the scattering pattern [27]. A laser beam illuminates the fiber normally, and the forward far-field scattering pattern is detected as a function of the scattering angle θ by a photomultiplier mounted on a rotatable arm. The fiber is immersed in matching oil whose refractive index is nearly equal to that of the fiber surface. Thus, the scatter at the fiber surface is almost removed, and the scatter due to the inner refractive-index variation can be effectively detected. The vessel of the matching oil has a plane glass window for the incident laser beam and a cylindrical window for the scattered light.

Because the necessary dynamic range for measuring the scattering pattern is very wide (typically 60 dB), a wide-dynamic-range photomultiplier is

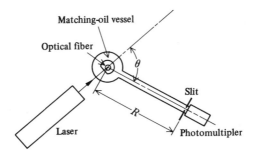

Fig. 9.1   Optical setup for measuring the scattering pattern.

used as the detector; moreover, when necessary, the power level of the incident beam is switched by using an optical attenuator. A narrow vertical slit is placed in front of the photomultiplier to improve the horizontal resolution, still picking up all the light energy extending in the vertical direction.

9.2.2   Derivation of Formula for Computing the Refractive-Index Profile

Before deriving the formula for computing the refractive-index profile from the scattering pattern, we must first derive the "forward" formula, i.e., the formula giving the scattering pattern in terms of the index profile. The latter formula can be derived by using the induced dipole method described in Appendix 9A.1; in this method the scattered field is computed by considering distributed equivalent dipoles corresponding to the spatial variation of the refractive index.

The coordinates and symbols used in the analysis are shown in Fig. 9.2. The refractive-index distribution is assumed to be axially symmetric. Point P

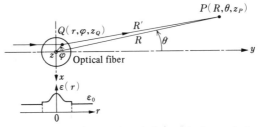

Fig. 9.2   Coordinates and symbols used in the analysis.

in Fig. 9.2 represents a point in the detector plane. The fiber to be tested is placed along the z direction. The laser beam is assumed to be polarized in the same z direction.

Let the permittivity of the matching oil, the permittivity profile inside the fiber, and the amplitude of the incident laser beam be denoted by ε_0, $[\varepsilon_0 + \varepsilon_1(r)]$, and A_0, respectively, where r is the radial coordinate. By using the induced dipole method, the scattered field dE_s produced in a small volume dV surrounding a point Q in the fiber is expressed as

$$dE_s = -\frac{k_0^2}{4\pi} \frac{\varepsilon_1(r)}{\varepsilon_0} \frac{1}{R'} A_0 e^{-j\Phi} e^{j\omega t} \, dV, \tag{9.1}$$

where k_0 is the wave number in the matching oil and Φ the phase of the scattered light.

We next make the following three assumptions.

(1)   The incident beam amplitude A_0 is constant everywhere in the fiber.
(2)   Both the incident beam and the scattered light propagate straight in the fiber; i.e., the refraction is negligible.
(3)   The additional phase shift caused by $\varepsilon_1(r)$ is negligible.

Then we may write the phase shift Φ approximately as

$$\Phi = k_0 R - 2k_0 r \sin(\tfrac{1}{2}\theta)\{\sin(\varphi - \tfrac{1}{2}\theta)\} + (k_0/2R)(z_P - z_Q)^2, \tag{9.2}$$

and the scattered field E_s becomes [27]

$$\begin{aligned}
E_s &\simeq \frac{k_0^2 A_0}{4\pi R} \int_{-l/2}^{l/2} \int_0^{2\pi} \int_0^{\infty} \frac{\varepsilon_1(r)}{\varepsilon_0} r \exp\{-j\Phi\} \, dr \, d\varphi \, dz_Q \\
&= \frac{k_0^2 A_0}{4\pi R} \int_0^{\infty} \frac{\varepsilon_1(r)}{\varepsilon_0} r \left[\int_0^{2\pi} \exp\{j2k_0 \sin(\tfrac{1}{2}\theta) \cdot r \sin(\varphi - \tfrac{1}{2}\theta)\} \, d\varphi \right] dr \\
&\quad \times l \operatorname{sinc} \frac{k_0 l z_P}{2R} \\
&= \frac{k_0^2 A_0}{2R} l \operatorname{sinc} \frac{k_0 l z_P}{2R} \int_0^{\infty} \frac{\varepsilon_1(r)}{\varepsilon_0} r J_0(2k_0 \sin(\tfrac{1}{2}\theta) \cdot r) \, dr. \tag{9.3}
\end{aligned}$$

In this equation, l denotes the width of the incident beam in the z direction, J_0 the zeroth-order Bessel function of the first kind, and $\operatorname{sinc} x \equiv \sin x / x$.

We define here the normalized scattered power σ as $\sigma = 2\pi R P_s / P_i$, where P_s denotes the scattered light power entering the detector and P_i the incident beam power measured by using the same detector. From Eq. (9.3),

$$\text{sign}[E_s(k_t)] \sqrt{\frac{1}{k_0^3 \pi^2}} \sqrt{\sigma(k_t)} = \int_0^\infty \frac{\varepsilon_1(r)}{\varepsilon_0} r J_0(rk_t)\, dr, \qquad (9.4)$$

where $\text{sign}[E_s(k_t)]$ denotes the polarity of the scattered field to be discussed in detail in Section 9.2.3, and

$$k_t = 2k_0 \sin(\theta/2). \qquad (9.5)$$

The right-hand side of Eq. (9.4) is the Hankel transform of $\varepsilon_1(r)/\varepsilon_0$. Consequently, by taking the inverse Hankel transform (which is also a Hankel transform [31]) of Eq. (9.4), the relative permittivity $\varepsilon_1(r)/\varepsilon_0$ is obtained as

$$\frac{\varepsilon_1(r)}{\varepsilon_0} = \frac{\sqrt{k_0}}{\pi} \int_0^\infty \text{sign}[E_s(\theta)] \sqrt{\sigma(\theta)} J_0(2k_0 \sin(\tfrac{1}{2}\theta) \cdot r) \sin\theta\, d\theta. \qquad (9.6)$$

The refractive-index profile $n(r)$ can be obtained from $\varepsilon_1(r)/\varepsilon_0$ as

$$n(r) = n_0\{1 + \varepsilon_1(r)/\varepsilon_0\}^{1/2}, \qquad (9.7)$$

where n_0 is the reflective index of the matching oil.

9.2.3   Determination of the Polarity of the Scattered Field

Equation (9.6) suggests that the determination of the polarity $\text{sign}[E_s(\theta)]$ is required for computing $\varepsilon_1(r)$. At angles where $\text{sign}[E_s(\theta)]$ changes, $|E_s(\theta)|$ should theoretically be zero because of its continuity with θ. Consequently, at least in principle, we may consider that at those angles at which the scattered power approaches zero the polarity of the scattered field changes (see region A in Fig. 9.3). However, because of the wide dynamic range of

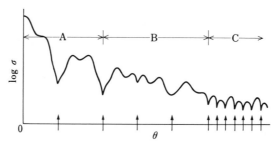

Fig. 9.3   Typical scattering pattern showing angles at which the polarity of the scattered field changes (vertical arrows).

the scattering pattern and the optical noise superposed, detection of the polarity change by this scheme often fails.

The polarity change can be detected more effectively by intentionally leaving a little refractive-index mismatch between the fiber surface and the matching oil as shown in Fig. 9.2 (lower figure), thus generating a weak scatter at the fiber surface. In such a case the measured scattering pattern consists of two components, i.e., a component caused by the refractive-index variation in the core, and a ripple caused by the scatter at the fiber surface. The former varies more slowly than the latter, whereas the amplitude of the latter is relatively small, as illustrated in Fig. 9.3. We can detect the polarity change from the intensionally introduced small ripple as shown in region B of Fig. 9.3, because the ripple phase is inverted at those points at which the polarity change occurs (vertical arrows).

When the index variation at the core–cladding boundary is smooth, the polarity of the ripple changes twice as fast at large scattering angles where the surface scattering predominates as shown in region C of Fig. 9.3. In such a case, sign$[E_s(\theta)]$ should change the polarity at each minimum of the scattering pattern [27]. We can also determine the outer diameter of the cladding from the period of minima in this region. In cases of single-mode fibers, the entire scattering pattern often becomes like region C of Fig. 9.3 because the scattering from the core is very weak.

9.2.4 Accuracy

The induced dipole method is usable only for small perturbations. There-fore, we should investigate the limits of core radius and permittivity variation above which the error in the computed profile becomes intolerable.

The rigorous expression for the scattering pattern of a uniform-core fiber immersed in matching oil having the same refractive index as the cladding is [32]

$$\sqrt{\sigma(\theta)} = \text{sign}\left\{\sqrt{\sigma(\theta)}\right\} (4/k_0) \left| b_0 + 2 \sum_{n=1}^{\infty} b_n \cos n\theta \right|, \tag{9.8}$$

where

$$b_n = \frac{mJ_n(\alpha)J_n'(m\alpha) - J_n'(\alpha)J_n(m\alpha)}{mH_n(\alpha)J_n'(m\alpha) - H_n'(\alpha)J_n(m\alpha)} \tag{9.9}$$

and $\alpha = k_0 r_{core}$ (r_{core} is the core radius), $m = n_1/n_2$ (n_1 is the refractive index of the core, and n_2 the refractive index of the cladding), and $H_n(z)$ is the nth-order Hankel function of the second kind.

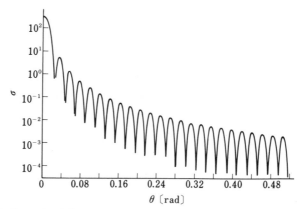

Fig. 9.4   Example of the normalized scattering pattern computed using Eq. (9.8).

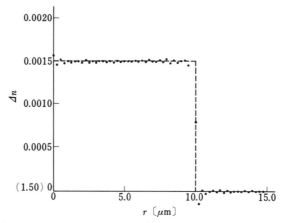

Fig. 9.5   Refractive-index profile computed using Eqs. (9.6) and (9.7) and the scattering pattern shown in Fig. 9.4. --- correct solution; • profile computed from the scattering pattern.

Figure 9.4 shows the normalized scattering power computed by using Eq. (9.8), for a uniform-core fiber having a 10-μm core radius and a 0.1% refractive-index difference. Dots in Fig. 9.5 show the profile calculated from Fig. 9.4 using Eqs. (9.6) and (9.7). In this case the sampling interval of the scattering pattern and the maximum sampling angle are 0.001 and 1.0 radian, respectively. In Fig. 9.5 the assumed profile is also shown by dashed lines, with which the computed profile (dots) shows a fairly good agreement but with a certain amount of systematic error.

As a measure of the computation error of the profile, we tentatively define a parameter

$$E = \langle |n - n'| \rangle / \Delta n, \qquad (9.10)$$

where $\langle \ \rangle$, n, n', and Δn denote an average over the core radius, the computed refractive index, the given refractive index, and the given refractive-index difference between the core and cladding, respectively.

Figure 9.6 shows the error E obtained with various simulations as functions of the product $r_{core} \times \Delta n/n$. It is found that the error is principally determined by this product. If we allow an error of 5%, the present method is applicable without correction to fibers whose $r_{core} \times \Delta n/n$ product is smaller than 0.04 (μm). Thus, generally, this method shows good accuracy for fibers having a relatively small number of propagation modes.

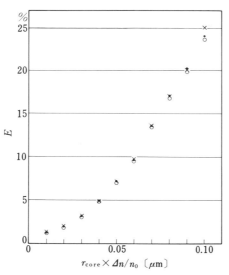

Fig. 9.6 The error E obtained with various simulations as functions of $r_{core} \times \Delta n/n$, $r_{core} = 5 \ \mu m \ (\times)$, $= 10 \ \mu m \ (\cdot)$, $= 20 \ \mu m \ (\cdot)$.

9.2.5 Spatial Resolution

In computing the index profile from the scattering pattern, the data desirably should be collected for $k_t = 0 - \infty$. However, data for k_t greater than

$$k_{max} = 2k_0 \qquad \text{(corresponding to } \theta = \pi) \qquad (9.11)$$

can never actually be obtained. In actual measurements, due to the limited dynamic range of the detector, a smaller effective range k_m exists for which

$$k_m < k_{\max}, \tag{9.12}$$

and the refractive-index profile must be computed from a truncated scattering pattern inside $\pm k_m$, i.e., we must assume that, for $k_m < k_t$,

$$\sqrt{\sigma(k_t)} = 0. \tag{9.13}$$

In such a case, by using the sampling theorem for the Hankel transform described in Appendix 9A.2, it is shown that the truncated scattering pattern as shown by Eq. (9.13) can be determined uniquely by refractive indices at discrete radii:

$$r_l = \alpha_{0l}/k_m, \tag{9.14}$$

where α_{0l} is the lth zero of the zeroth-order Bessel function. Consequently, the average spatial resolution is

$$\Delta r = \langle r_{l+1} - r_l \rangle \simeq \pi/k_m, \tag{9.15}$$

where the symbol $\langle \; \rangle$ denotes an average over k.

Because the maximum value of k_m is $2k_0$, the resolution limit of this method is given as

$$\Delta r_{\min} = \pi/2k_0 = \lambda_0/4. \tag{9.16}$$

Thus, the resolution limit of this method is equal to the intrinsic limitation of the optical method in general.

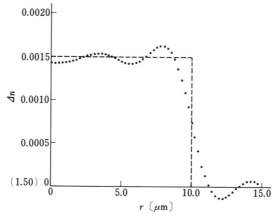

Fig. 9.7   Example of the profile obtained when the range of the scattering pattern is not wide enough.

Figure 9.7 shows a scattering-pattern profile where the scan range is too small; in this case, 0–0.1 radian. According to Eq. (9.15), this range gives a spatial resolution of 2.1 μm, which is nearly equal to half the ripple period in Fig. 9.7. Thus, Eq. (9.15) is substantiated [27].

9.2.6 System of Measurement

In the scattering-pattern method, the sampling interval of the scattering pattern has to be small to determine the lobe polarity from the small ripple caused at the fiber surface, whereas the angular range of the measurement has to be wide to get high resolution. Thus, a typical number of the sampling points is 1000–2000.

To collect and process such a large number of data efficiently, an automated system as shown in Fig. 9.8 must be used. The photomultiplier is mounted on a rotatable arm which is motor-driven at constant speed. The outputs of the photomultiplier and the angular potentiometer are amplified and fed into a minicomputer through A/D (analog-to-digital) converters and

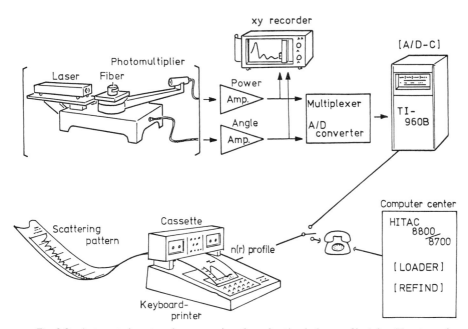

Fig. 9.8 Automated system for measuring the refractive-index profile (after Hotate and Okoshi [33]).

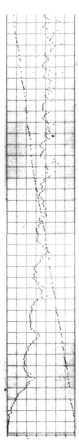

Fig. 9.9 Part of the measured scattering pattern for a uniform-core fiber with $r_{core} = 10$ μm and $\Delta = 0.0037$. The ordinate shows $\log[\sigma(\theta)]$, and the sampling interval in the abscissa is 0.586×10^{-4} radian. The maximum sampling angle is 0.698 radian; only one-third of the pattern (0–0.22 radian) is shown in this figure (graphic line-printer output). (After Okoshi and Hotate [27].)

multiplexer. The collected and preprocessed data are sent to the computer via a TSS terminal for the analysis.

9.2.7   Result of Measurement

Figure 9.9 shows an example of the measured scattering pattern $\sigma(\theta)$ for a uniform-core fiber with design values $r_{core} = 10$ μm and $\Delta n = 0.0037$.

Figure 9.10 shows the profile computed by using the scattering pattern of Fig. 9.9 from 0 to 0.698 radian [27]. The theoretical spatial resolution given by Eq. (9.15) is 0.3 μm.

Figure 9.11 shows an example of refractive-index profiles obtained for single-mode fibers [33]. The scattering pattern was measured for the angular

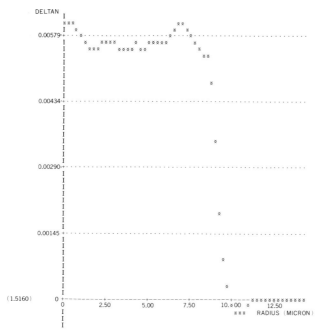

Fig. 9.10 Refractive-index profile computed from the scattering pattern shown in Fig. 9.9 (graphic line-printer output) (after Okoshi and Hotate [27]).

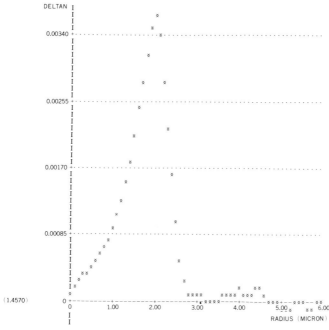

Fig. 9.11 Refractive-index profile obtained for a single-mode fiber (after Hotate and Okoshi [33]).

range 0–1.24 radian, which gives a spatial resolution of 0.19 μm. The single-mode fiber under test is called O-type for its cylindrical core.

9.2.8   Other Methods Using the Scattering Pattern

Two methods will be described, which are also based on the scattering-pattern data, but are usable only for restricted profiles.

The first method is coded A in Table 9.1. In this method, the refractive index and radius of a uniform "unclad" fiber can be obtained by using the optical setup shown in Fig. 9.12 [1].

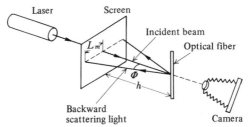

Fig. 9.12   Optical setup for measuring the backward scattering pattern (after Presby [1]).

We consider a ray that is internally reflected at the surface of an unclad fiber and travels backwards, as shown in Fig. 9.13. As will be seen, a scattering angle Φ_m exists for which the scattered light intensity becomes maximum. This angle Φ_m is determined solely by the refractive index of the fiber n; hence, the index can be determined by measuring Φ_m.

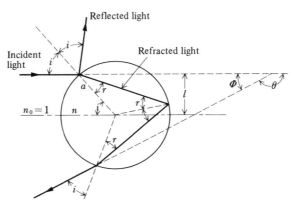

Fig. 9.13   Formation of the backward scattering light (after Presby [1]).

The relation between Φ_m and n is obtained as follows. From Fig. 9.13,

$$\theta = \pi + 2i - 4r. \tag{9.17}$$

On the other hand,

$$r = \sin^{-1}\{(\sin i)/n\}. \tag{9.18}$$

Hence, from the condition $d\theta/di = 0$, the incident angle i_m corresponding to Φ_m is

$$i_m = \cos^{-1}\{(n^2 - 1)/3\}^{1/2}. \tag{9.19}$$

On the other hand,

$$\Phi = 4r - 2i. \tag{9.20}$$

Therefore, from Eqs. (9.18)–(9.20), we obtain

$$\Phi_m = 4\sin^{-1}\{(2/n\sqrt{3})(1 - \tfrac{1}{4}n^2)^{1/2}\} - 2\sin^{-1}\{(2/\sqrt{3})(1 - \tfrac{1}{4}n^2)^{1/2}\}. \tag{9.21}$$

This is the equation giving the relation between Φ_m and n.

Fig. 9.14 Example of the backward scattering pattern (after Presby [1]).

Next, we consider a method used for obtaining the fiber radius. Figure 9.14 shows an example of the backward scattering pattern. This figure shows the presence of a ripple attributed to the interference between two rays ACDEG and HIJ shown in Fig. 9.15. The fiber radius can be determined from the period of this ripple.

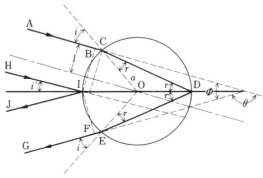

Fig. 9.15 Generation of the ripple in the backward scattering pattern (after Presby [1]).

The path-length difference between the two rays is, using symbols defined in Fig. 9.15,

$$S = BC + CD + DE + EF = 2(BC + CD), \tag{9.22}$$

where

$$BC = a\{\cos(2r - i) - \cos i\} \tag{9.23}$$

$$CD = 2na\cos r. \tag{9.24}$$

Therefore, from Eqs. (9.18), (9.22), (9.23), and (9.24), we obtain

$$S = 4a\left\{\left(1 - \frac{\sin^2 i}{n^2}\right)^{1/2}\left(n + \frac{\sin^2 i}{n}\right) - \frac{\sin^2 i \cos i}{n^2}\right\}. \tag{9.25}$$

If we expand Eq. (9.25) in a power series of i and collect up to the i^2 term, we can rewrite Eq. (9.25), using Eqs. (9.18) and (9.20), as a function of Φ,

$$S \simeq 4a\left\{n + \frac{\Phi^2}{16(1 - \tfrac{1}{2}n)}\right\}. \tag{9.26}$$

The difference between the path lengths at two adjacent minima (or maxima) of the scattering pattern (Fig. 9.14) must be equal to one wavelength λ. Therefore, if the corresponding scattering angles are denoted by Φ_1 and Φ_2, then

$$\lambda = 4a\left\{\frac{\Phi_1^2 - \Phi_2^2}{16(1 - \tfrac{1}{2}n)}\right\}. \tag{9.27}$$

On the other hand, Fig. 9.12 suggests that

$$\Phi \simeq L/h. \tag{9.28}$$

Hence

$$\Delta L = L_1 - L_2 = (2\lambda h^2/aL)(1 - \tfrac{1}{2}n), \tag{9.29}$$

and the radius of the fiber is

$$a = [2\lambda h^2/L(\Delta L)](1 - \tfrac{1}{2}n). \tag{9.30}$$

By using the foregoing method, Presby succeeded in measuring the refractive index with an accuracy of 0.2%, and the core radius with an accuracy of 5%. [1]

The second method is coded B in Table 9.1. In this method, the optical setup shown in Fig. 9.12 is used, but parameters of uniform-core, cladded (two-layer) fibers can be determined [2].

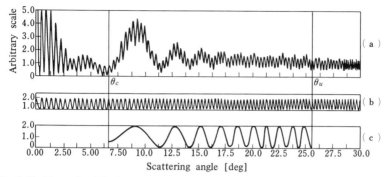

Fig. 9.16 Example of the forward scattering pattern of a uniform-core fiber (after Watkins [2]): (a) measured pattern; (b, c) based on geometrical optics.

An example of the measured scattering pattern of a uniform-core cladded fiber is shown in Fig. 9.16. This pattern consists of a fine ripplelike component carrying the cladding information, and a slowly varying component carrying the core information. The intensity distributions of these two components can be computed from geometrical optics by computing the phase difference between two sets of rays (one set consisting of two rays reflecting at the cladding-outside boundary, and the other set consisting of two rays reflecting at the core–cladding boundary), in a manner similar to obtaining Eq. (9.26). Examples of such computations are shown in Figs. 9.16b and 9.16c. In the region $\theta_u < \theta$, only the fine ripplelike component is present; the radius of the cladding can be determined from this portion of the scattering pattern provided the cladding index is known [2]. The radius and index of the core are determined, using the least-squares method, so that the computed and measured slowly varying components show the best fit.

The drawback of this method is that in actual uniform-core cladded fibers, the boundary between the core and cladding becomes more or less gradual due to the thermal diffusion of dopants during the drawing, making the determination of the parameters difficult.

9.3 Interference Methods

9.3.1 Principle

We consider a Mach–Zehnder interferometer shown in Fig. 9.17. When the incident light beam is a plane wave and the two wavefronts reflected by M_1 and M_2 are exactly parallel after they are merged by HM_2, only uniform

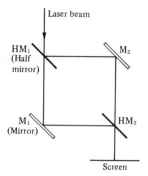

Fig. 9.17   A Mach–Zehnder interferometer.

Fig. 9.18   Parallel interference fringes obtained for two plane waves.

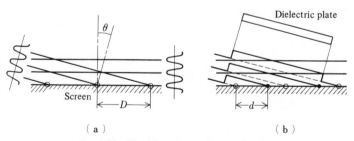

Fig. 9.19   The fringe-formation mechanism.

irradiation will be observed on the screen. When the two wavefronts make a small angle, parallel interference fringes as shown in Fig. 9.18 will be observed.

The fringe-formation mechanism is illustrated in Fig. 9.19a. We consider two wavefronts tilted by θ with each other, and assume that one is normally incident on the screen. Two sets of parallel solid lines in Fig. 9.19a show such wavefronts. The small circles indicate the positions at which the two wavefronts enhance each other and hence the brightness increases. Between these circles, the two waves cancel and the brightness is reduced. Thus, the distance between two adjacent fringes corresponds to the relative phase difference of 2π.

Next we assume that a small dielectric plate is inserted in the tilted path as shown in Fig. 9.19b. In this case the enhanced point moves a distance d, from the small circle to the dot shown in Fig. 9.19b. In such a case the additional phase shift in the dielectric plate is expressed in terms of d as

$$\phi = 2\pi d/D, \qquad (9.31)$$

where D denotes the spacing between fringes (see Fig. 9.19a).

The principle of the interference method is to determine the refractive-index distribution in an object from the lateral shift of the interference fringe, which gives the phase shift of light passing through it.

9.3.2 Microscopic Interferometry Using Sliced Samples

Figure 9.20 shows an optical setup for measuring the refractive-index profile of an optical fiber using an interference microscope. The fiber to be tested is first molded in epoxy resin, and then sliced normal to the axis of the fiber to form a thin cross-sectional sample having a thickness 0.1–0.5 mm. The surface of the sliced sample is polished, and then the sample is inserted into the microscope. In such a case an interference-fringe pattern as shown

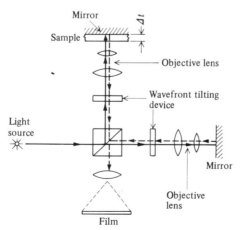

Fig. 9.20 Optical setup for measuring the index profile of fibers using an interference microscope (after Martin [8]).

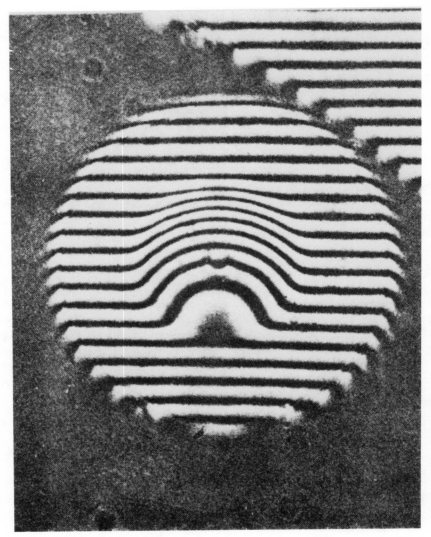

Fig. 9.21   Example of the interference-fringe pattern (after Wonsiewicz *et al.* [10]).

in Fig. 9.21 is observed; the index profile of the fiber can be computed from such a pattern in the following way [10].

We first denote the index distribution

$$n(x, y) = n_2 + \Delta n(x, y), \tag{9.32}$$

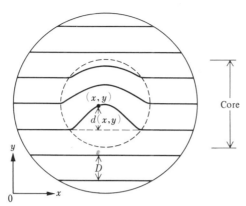

Fig. 9.22 Definition of the fringe shift $d(x, y)$ in terms of the coordinate (x, y).

where n_2 is the cladding index. The phase shift due to $\Delta n(x, y)$ is, using the wave number in vacuum k and the sample thickness t,

$$\phi = k \, \Delta n(x, y) 2t. \tag{9.33}$$

Hence, from Eqs. (9.31) and (9.33), $n(x, y)$ is expressed, by using the fringe shift $d(x, y)$ shown in Fig. 9.22, as

$$\Delta n(x, y) = [d(x, y)/D](\lambda/2t). \tag{9.34}$$

Note that $d(x, y)$ is defined in terms of the point (x, y) shown as a dot on the fringe. When a transmission-type interference microscope is used, the factor 2 in the denominator of Eq. (9.34) should be deleted because the ray passes through the sample only once.

Wonsiewicz *et al.* [10] developed an automated system for measuring the index profile by combining an interference microscope, scanning densitometer, AD (analog-to-digital) converter, and a computer. An example of the obtained profile is shown in Fig. 9.23; this has been computed from the interference fringe pattern shown in Fig. 9.21. In this measuring system, the spatial resolution and accuracy are $0.7 \, \mu m$ and $\pm 5 \times 10^{-4}$, respectively [10].

In spite of the tedious sample preparation required, this method is widely used in factories and laboratories and in fact, the optical system is commercially available. Major error sources are the effect of the ray bending (especially when the sample is thick) and the roughness of the sample surface. The accuracy is reduced at places where the index changes abruptly. For very thin fibers such as single-mode fibers, precise data cannot be obtained because of the limited spatial resolution,

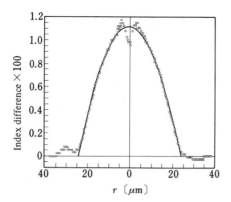

Fig. 9.23   Example of the profile obtained (after Wonsiewicz [10]).

9.3.3   Microscopic Interferometry Using As-Drawn Samples

The method just described is universal, because it is applicable to any profile. However, it is a destructive method. A nondestructive version will now be described.

We can obtain the interference-fringe pattern of a fiber in a nondestructive manner, if we put the sample into one of the optical paths of a Mach–Zehnder interferometer so that the ray passes the fiber normally as shown in Fig. 9.24. In this case, however, the fiber must be immersed in matching oil whose refractive index is nearly equal to that at the fiber surface. In such a case we can consider that the ray passes through the fiber almost straight,

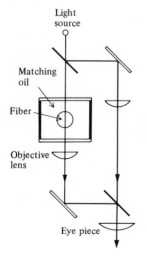

Fig. 9.24   Microscopic interferometry using as-drawn samples.

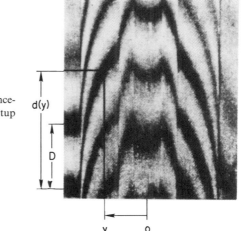

Fig. 9.25 Example of the interference-fringe pattern obtained with the optical setup shown in Fig. 9.24 (after Shiraishi [21]).

and hence compute the phase shift in the fiber. An example of the inter-ference-fringe pattern thus obtained is shown in Fig. 9.25 [21].

The index profile can be computed from the interference-fringe pattern in the following way. We assume for simplicity that the fiber is axially symmetrical and that the rays travel straight through the fiber as shown in Fig. 9.26. Under such assumptions, the excess phase shift is given in terms of the y coordinate as

$$\phi(y) = 2k \int_y^R \frac{\Delta n_m(r) r}{(r^2 - y^2)^{1/2}} \, dr, \qquad (9.35)$$

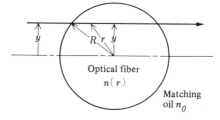

Fig. 9.26 Computation of the phase shift of rays.

where $\Delta n_m(r) = n(r) - n_0$ (n_0 is the index of matching oil), and R denotes the cladding radius. Substituting Eq. (9.35) into (9.31), we obtain

$$d(y) = \frac{2D}{\lambda} \int_y^R \frac{\Delta n_m(r) r}{(r^2 - y^2)^{1/2}} \, dr. \qquad (9.36)$$

This equation has the form of Abel's integral equation, and can readily be solved for $\Delta n_m(r)$ to give [22]

$$\Delta n_m(r) = -\frac{\lambda}{\pi D} \int_r^R \frac{d[d(y)]}{dy} \frac{dy}{(y^2 - r^2)^{1/2}}. \tag{9.37}$$

In the derivation of this equation, the straight path of the ray is assumed. Therefore, when the index variation in the fiber is relatively abrupt, the error due to the ray bending increases. Therefore, from this viewpoint this method is more suitable for low-Δ fibers. However, we should also note that in such cases the relative fringe shift $d(y)/D$ decreases; this degrades the overall accuracy of the measurement.

9.3.4   Differential Interferometry

Suppose we slightly modify the Mach–Zehnder interferometer as shown in Fig. 9.27, and observe the interference-fringe pattern between two beams which both pass the fiber. One beam is laterally shifted a small distance s by using a shearing device built in the interferometric microscope. In such a case, we observe the interference-fringe pattern corresponding to the derivative of $d(y)$, as will be shown in the following. This method is called differential interferometry [23, 34].

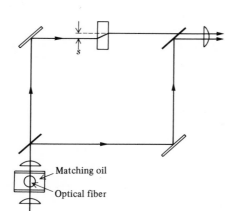

Fig. 9.27   Optical setup for the differential interferometry (after Iga *et al.* [23]).

In this case what appears as the fringe shift is the phase difference between two rays passing the fiber at y and $y + s$ in Fig. 9.28. This phase difference can be computed simply by using $\phi(y)$ obtained in Eq. (9.35),

$$\phi_s = \phi(y + s) - \phi(y), \tag{9.38}$$

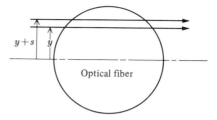

Fig. 9.28   Two rays contributing to the formation of the differential interferogram.

which becomes for a small value of s

$$\phi_s = \lim_{s \to 0} \frac{\phi(y + s) - \phi(y)}{s} s = \frac{d\phi(y)}{dy} s. \tag{9.39}$$

Therefore, the fringe shift is

$$d_s(y) = \frac{D}{2\pi} \frac{d\phi(y)}{dy} s. \tag{9.40}$$

Comparing Eqs. (9.31) and (9.40), we find that the fringe shift obtained with ordinary interferometry $d(y)$ and the preceding $d_s(y)$ are related by

$$d_s(y) = \frac{d[d(y)]}{dy} s. \tag{9.41}$$

This equation gives the required relation. The index distribution is given in this case more directly than in Eq. (9.37) as

$$\Delta n_m(r) = -\frac{\lambda}{\pi D s} \int_r^R d_s(y) \frac{dy}{(y^2 - r^2)^{1/2}}, \tag{9.42}$$

in which computation of the derivative is no longer necessary. This is surely a great advantage in processing the data; however, in this method it is sometimes rather difficult to obtain a clear interference pattern.

9.3.5   Interferometry for Preforms

In many fiber-manufacturing processes such as the MCVD method described in Section 2.11.3, the fiber material is once formed in a rod having the index profile to be obtained in the fiber. This rod, typically 1–3 cm in diameter and 100 cm long, is called a preform rod or simply a preform. The preform is then put into a furnace and the fiber is drawn from it.

In such a manufacturing process, the precise measurement of the preform index is as important as, or even more important than, the index profile measurement of the drawn fiber. This is so because the index profiles before and after drawing are almost similar, and hence we can discard bad preforms (if there are any) to improve the utility of the drawing machine.

To measure a preform index profile, the method described in Section 9.3.2 (interferometry with sliced sample) is used most commonly. This method is easier and more accurate for such a large sample than for a drawn fiber having a tiny cross section. Every statement on the advantages and disadvantages of this method described in Section 9.3.2 also remain valid for preforms.

An alternative, nondestructive method will now be described. This method is based on the same principle as that described in Section 9.3.3. To minimize the error due to ray bending, we use the optical setup shown in Fig. 9.29, in which the interference-fringe pattern on the median plane of the preform (shown as a dashed vertical line) is focused on the screen by a convex lens. This scheme is called the "fringe localization," and is commonly used in optical interferometry. (Actually, we employ this scheme unconsciously also in the fiber measurement shown in Fig. 9.20, by adjusting the microscope to the best fringe pattern.)

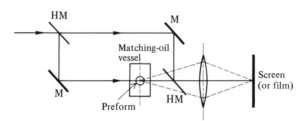

Fig. 9.29 Mach–Zehnder interferometer using the fringe-localization scheme for measuring the preform index profile.

When the fringe-localization scheme is employed, the ray-bending correction can be made without difficulty. For example, the measurement can be performed with satisfactory accuracy for a typical preform having a 5-mm core diameter and a 1% relative index difference [30]. Because of the large preform diameter, the microscope is usually not required.

If we assume that the indices of the matching oil and the preform cladding are exactly equal, the fringe shift obtained with the setup shown in Fig. 9.29 can be computed from the phase difference between two rays $P_1P_2P_3P_4$ and $P_5P_6P_7P_4$ shown in Fig. 9.30,

$$\phi_t = P_P(\widehat{P_1P_2P_3P_4}) - P_R(P_5P_6P_7P_4). \tag{9.43}$$

In this equation, P_P and P_R denote the phase shifts of the ray passing through the preform and the reference beam, respectively. On the other hand,

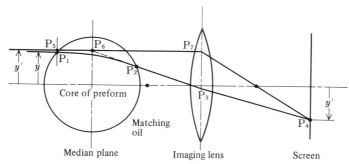

Fig. 9.30   Computation of the phase shift of rays in the optical system shown in Fig. 9.29 (after Okoshi *et al.* [30]).

from the property of a convex lens,

$$P_R(P_5P_6P_7P_4) = P_R(P_5P_6P_2P_3P_4). \tag{9.44}$$

Hence, we obtain

$$\phi_t = P_P(\widehat{P_1P_2}) - P_R(P_5P_6P_2). \tag{9.45}$$

This phase shift can be computed in terms of the index profile $n(r)$ without difficulty [22, 35, 36]. By a change of variable,

$$u = rn(r), \tag{9.46}$$

the fringe shift $d(y)$ corresponding to ϕ_t can be related to the function $n(u)$ by Abel's integral equation. Solving it, we obtain [31, 34]

$$n(u) = n_m \exp\left\{-\frac{\lambda}{\pi D}\int_u^{u_R}\frac{d[d(y)]}{d(yn_m)}\frac{d(yn_m)}{\sqrt{(yn_m)^2 - u^2}}\right\}, \tag{9.47}$$

where $u_R = r_c n_m$, with r_c denoting the core radius. This equation does not give the function $n(r)$ explicitly, but gives it only implicitly. The radius r corresponding to a value of $n(u)$ can be obtained as

$$r = u/n(u). \tag{9.48}$$

We should note that the foregoing scheme gives a correct result only when $u = rn(r)$ increases monotonously as a function of r.

9.3.6   Use of Interference between Lights Reflected at Both Ends of the Fiber

Suppose that a fiber has an index profile which is close to quadratic,

$$n(r) = \varepsilon_s(r)^{1/2} = \varepsilon_{s1}^{1/2}\{1 - (\alpha r)^2 + C_4(\alpha r)^4 + \cdots\}^{1/2}, \tag{9.49}$$

where $\varepsilon_s(r)$ and ε_{s1} are the permittivities in the core and on the axis, respectively. In some cases a very precise measurement of the coefficients of higher-order terms C_4, C_6, ... is required, because the coefficients have critical effects on the transmission characteristics of the fiber. Rawson *et al.* [6] developed a method for measuring these coefficients using interference between two lights reflected at both ends of the fiber under test.

The optical setup is shown in Fig. 9.31. When the fiber is nearly quadratic, a beam once focused on the input plane again converges at a certain distance as shown in Fig. 9.31. We denote this distance by L. [Strictly speaking, as described in Section 3.3.3, an index profile having a form $\mathrm{sech}(\pi r/L)$ is required for the pin-point focus.] If the fiber sample in Fig. 9.31 is of length L and has optically flat end surfaces (A and B), light reflected at both ends will return along almost the same paths and will make interference fringes on the film at the left end. The quarter-wavelength plate is inserted to remove unwanted components scattered in the prism or lens. When the incident light is polarized normal to the paper, it becomes circularly polarized by the quarter-wavelength plate, reflected at both ends, and again becomes linearly polarized in parallel with the paper after passing the plate. Thus, the unwanted components can be removed by the polarizer in front of the film.

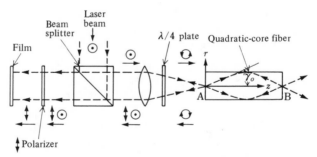

Fig. 9.31   Optical setup for preparing the interferogram between lights reflected at both ends of a fiber (after Rawson *et al.* [6]).

An example of the interferogram obtained by this method is shown in Fig. 9.32. Noting that two neighboring fringes give a phase difference 2π, we can directly obtain the phase difference between lights reflected at two ends of the sample as a function of the incident angle γ_0 shown in Fig. 9.31.

On the other hand, this phase difference can be computed on geometrical optics if we express the profile as Eq. (9.49). As described in Appendix 9A.3,

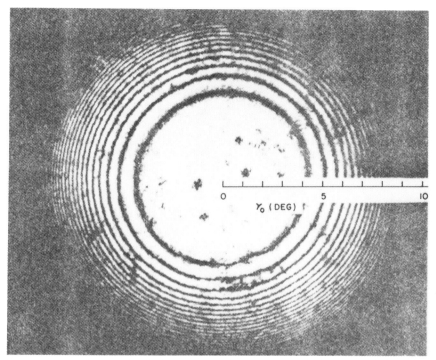

Fig. 9.32   Example of the interferogram between lights reflected at both ends of a fiber (after Rawson *et al.* [6]).

the path of a ray entering the fiber with angle γ_0 can be obtained by solving [37]

$$\frac{d^2 r}{dz^2} = \frac{1}{2\varepsilon_{s1}\cos^2\gamma_0}\frac{d\varepsilon_s(r)}{dr}. \tag{9.50}$$

On the other hand, the phase shift along the ray is

$$d\varphi = kn(r)(ds/dz)\,dz, \tag{9.51}$$

where k and s denote the wave number in vacuum and the length along the ray, respectively. Since $ds^2 = (dr/dz)^2\,dz^2 + dz^2$, Eqs. (9.50) and (9.51) lead to [6]

$$d\varphi/dz = k\varepsilon_s(r)/(n_1\cos\gamma_0), \tag{9.52}$$

where $n_1 = (\varepsilon_{s1})^{1/2}$.

If we assume the values of parameters α, C_4, and C_6 in Eq. (9.49), we can compute the phase difference between two lights as a function of the angle γ_0 by using Eqs. (9.50) and (9.52). Therefore, using the least squares method, we can determine those parameters so that the computed phase shift exhibits the best fit with the measured one. Figure 9.33 shows the result of such a curve fitting for the interferogram shown in Fig. 9.32; the circles and the solid curve show the measured and computed phase shifts, respectively. The parameters thus determined were $\alpha = 2.135 \pm 0.002 \text{ cm}^{-1}$, $C_4 = 1.36 \pm 0.05$, and $C_6 = -3.0 \pm 0.4$ [6].

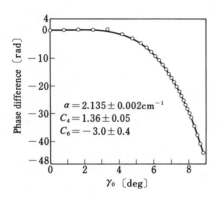

Fig. 9.33 Phase difference between two lights as a function of the angle γ_0, obtained for the interferogram shown in Fig. 9.32; $\alpha = 2.135 \pm 0.002 \text{ cm}^{-1}$, $C_4 = 1.36 \pm 0.05$, $C_6 = -3.0 \pm 0.4$ (after Rawson *et al.* [6]).

In this method, the error increases as the index profile deviates from $\text{sech}(\pi r/L)$. The sample must be prepared exactly with length L and both ends optically flat.

9.4 Reflection Method

9.4.1 Principle

When a ray of light is normally incident on a glass material having a refractive index n, the power reflection coefficient is, directly from Eq. (2.77),

$$R = [(1 - n)/(1 + n)]^2. \tag{9.53}$$

The principle of the reflection method [11–15] for measuring the index profile of an optical fiber rests entirely on Eq. (9.53).

An example of the optical setup is shown in Fig. 9.34. The laser beam is first purified by a simple mode filter (a tiny pinhole) and focused by an

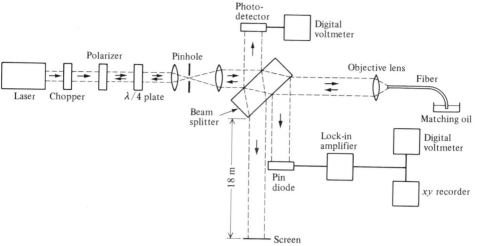

Fig. 9.34   Optical setup for the reflection method (after Eickoff *et al.* [13]).

objective lens on the end surface of the fiber being tested. Then the fiber is moved laterally and the reflection coefficient R is measured as a function of the lateral displacement of the fiber. From this result the index profile can be determined using Eq. (9.53). In the measuring system shown in Fig. 9.34, a chopper and a lock-in amplifier are used to improve the signal-to-noise ratio. The other end of the fiber is immersed in matching oil to remove the effects of reflection at this end.

We temporarily assume that the index difference between the core and cladding is much smaller than unity, as is the case in most practical fibers. Denoting the cladding index and reflection coefficient on cladding by n_2 and R_2, respectively, and letting

$$n = n_2 + \Delta n, \tag{9.54}$$

$$R = R_2 + \Delta R, \tag{9.55}$$

we may write

$$\Delta n \simeq \tfrac{1}{4}(n_2^2 - 1)(\Delta R/R_2). \tag{9.56}$$

This equation suggests that the index variation Δn is approximately proportional to the reflection-coefficient variation ΔR. Hence, the profile can be plotted directly on an xy recorder as shown in Fig. 9.34.

The spatial resolution is limited by the waist diameter of the incident beam; for an He–Ne laser generating a 632.8-nm wavelength, the resolution limit is 1.5–3 μm. In Ikeda *et al.* [11], Ueno and Aoki [12], and Eickhoff and Weidel [13], measuring systems with an overall accuracy of 5% are reported.

Because of relatively poor resolution, this method is not applicable to single-mode fibers. For multimode fibers, however, this method is sometimes used because of the simple optical setup and ease of sample preparation.

9.4.2   Examples of Measured Profile

Figures 9.35 and 9.36 show the results of measurement using the reflection method for a graded-core fiber made by the double-crucible method (Fig. 9.35) and that made by the MCVD method (Fig. 9.36) [12]. The dip at the center of Fig. 9.36 is due to the out-diffusion of the dopant during the collapse process (see Section 2.11.3).

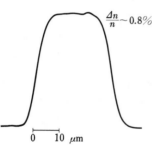

Fig. 9.35   Example of the index profile obtained using the reflection method—a fiber made by the double-crucible method (after Ueno *et al.* [12]).

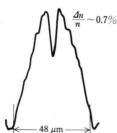

Fig. 9.36   Example of the index profile obtained using the reflection method—a fiber made by the MCVD method (after Ueno *et al.* [12]).

9.4.3   Improvement of Resolution

The spatial resolution limit in the reflection method is 1.5–3 μm. This value is satisfactory neither for single-mode nor for multimode fibers. To overcome this limitation, Shiraishi *et al.* [14] proposed the preparation of a relatively thick sample (having a diameter of 0.3 to 0.6 mm) from the tapered portion of the fiber produced at the end of drawing, and called it the *Q* fiber. By performing the reflection-method measurement with the *Q* fiber, they showed that an effective spatial resolution of 0.5 *μ*m could easily be achieved [14].

9.4.4 Vibrated Reflection Method

Aoki *et al.* [15] devised a method for improving the sensitivity of the reflection measurement by vibrating the fiber laterally. The essential part of the system is shown in Fig. 9.37. The reflected light is picked up, amplified, and detected synchronously with the vibrating signal. Thus, the spatial derivative of the index profile is obtained first, and the profile is finally obtained by integrating the signal using an operational amplifier.

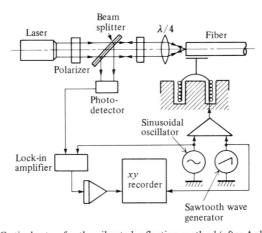

Fig. 9.37 Optical setup for the vibrated reflection method (after Aoki *et al.* [15]).

If the reflection coefficient at position *x* is denoted by $R(x)$, and *x* vibrates around a fixed point X as

$$x = X + \Delta x \cos \omega t, \tag{9.57}$$

then we may write

$$R(x) = R(X) + \Delta x R'(X) \cos \omega t + \cdots. \tag{9.58}$$

Therefore, if we pick up only the ω component, we will obtain $R'(X)$. In this method, the sensitivity is improved, but the spatial resolution is degraded at least by the width of vibration.

9.5   Other Methods

9.5.1   Near-Field Pattern Method

This is an approximate but very simple method for measuring the index profile of multimode fibers.

We consider a fiber having a profile $n(r)$. The wave number at the radial coordinate r,

$$k(r) = 2\pi n(r)/\lambda, \tag{9.59}$$

is related to the propagation constants in the three spatial directions in a manner shown in Fig. 9.38. Therefore, if the angle made by the wave vector and the z axis is denoted by $\theta(r)$ as shown in Fig. 9.38, the axial propagation constant β and the wave number $k(r)$ are related by

$$\cos \theta(r) = \beta/k(r). \tag{9.60}$$

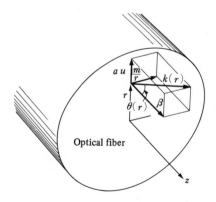

Fig. 9.38   Wave-number vector $k(r)$ and its three orthogonal components.

On the other hand, the cutoff condition is

$$\beta_2 = 2\pi n_2/\lambda, \tag{9.61}$$

where n_2 is the cladding index. Hence, it is found that, among those rays that enter the fiber at r, only those whose initial skew angle is smaller than θ_c

determined by

$$\cos \theta_c(r) = \beta_2/k(r) = n_2/n(r) \tag{9.62}$$

can propagage in the fiber. In other words, if we define the "local" numerical aperture (see Section 3.2.3) $A(r)$ in terms of the acceptable angle determined by Eq. (9.62), then

$$A(r) = n(r)\sin \theta_c(r) = \{n^2(r) - n_2^2\}^{1/2}. \tag{9.63}$$

We now assume that one end of the fiber is illuminated uniformly by incoherent light whose incident energy per unit volume angle is also uniform; thus, all propagating modes are excited uniformly. In such a case the power entering at r and propagating over certain distance can be expressed as [16]

$$P(r) = P(0)\frac{A^2(r)}{A^2(0)} = P(0)\frac{n^2(r) - n_2^2}{n^2(0) - n_2^2}. \tag{9.64}$$

If all the modes propagate with equal loss and mode conversion is absent, then the same power distribution as Eq. (9.64) will appear at the output end of the fiber. Therefore, if we write

$$n(r) = n_2 + \Delta n(r), \tag{9.65}$$

the near-field pattern at the output end can be expressed as

$$P(r) \simeq \frac{2P(0)n_2}{n^2(0) - n_2^2}\Delta n(r) \tag{9.66}$$

and is approximately proportional to the excess refractive index $\Delta n(r)$.

Figure 9.39 shows microscopic photograph of the near-field pattern at the end of a uniform-core multimode fiber [38]. The overall distribution of the index can be obtained conveniently by using this method. However, this method should be regarded as an approximate one. The reasons are: first, the difficulty in checking the input conditions (the uniformity in mode excitation), and second, the difference in attenuation for each mode. The second problem could be solved by making the sample shorter. However, the effect of the residual leaky modes then increases. Adams *et al.* [17] showed that this effect enhances the brightness at the peripheral part of the core, and proposed a simple corrected formula to replace Eq. (9.66):

$$P(r) = P(0)\frac{n^2(r) - n_2^2}{n^2(0) - n_2^2}\frac{1}{\sqrt{1 - (r/a)^2}}, \tag{9.67}$$

where a denotes the core radius.

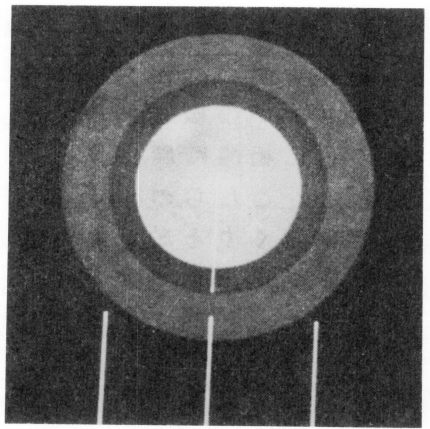

Fig. 9.39 Microscopic photograph of the near-field pattern at the end of a uniform-core fiber (after Inada *et al.* [38]).

9.5.2　X-Ray Microanalysis

In optical fibers for communication proposes, the index variation is provided by various dopants. It has been confirmed experimentally that the refractive-index variation is expressed approximately as the linear combination of the dopant densities [20]. On the other hand, the impurity densities can be measured effectively using the X-ray microanalyzer, because the intensity of the specific X-ray generated by an impurity material is proportional to the density of that impurity. Thus, we can use the X-ray microanalysis to determine the index profile. Since the exciting electron beam can be focused on a very small spot, the spatial resolution is very high. The practical limitation is a relatively poor signal-to-noise ratio.

9.5.3 Measurement of Higher-Order Terms in Nearly Quadratic Profile

An example of the method for measuring parameters representing higher-order terms in a nearly quadratic profile has been described in Section 9.3.6. Several other methods are listed in Table 9.1 (see methods D, E, and G). The details are omitted here because of space limitations.

9.5.4 Matching-Oil Method

In the matching-oil method, parameters of a cladded uniform-core fiber can be determined by immersing the fiber to be tested in matching oil whose refractive index can be controlled, and observing it through a microscope [3].

First, the cladding index can be determined from the fact that the boundary between the cladding and the matching oil becomes invisible when their indices coincide.

Next, the core index can be determined from the condition under which the boundary between the core and cladding becomes the least visible. This condition is illustrated in Fig. 9.40; it is given as the condition that the phase

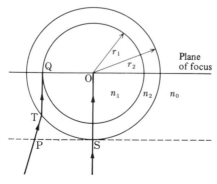

Fig. 9.40 Rays passing through a uniform-core fiber immersed in matching oil: $n_0 > n_1 \gtrsim n_2$ (after Liu [3]).

shifts along paths PTQ and SO are equal to each other. Hence,

$$\overline{PT} \cdot n_0 + \overline{TQ} \cdot n_2 = (r_2 - r_1)n_2 + r_1 n_1. \tag{9.68}$$

Therefore, the core index is

$$
\begin{aligned}
n_1 = (n_2/r)\{(1 - r^2)^{1/2} - (1 - r)\} \\
+ (n_0/r)\{1 - (1 - r^2)^{1/2}\}/[\{(1 - r^2)(1 - n^2 r^2)\}^{1/2} + nr^2], \tag{9.69}
\end{aligned}
$$

where $n = n_2/n_0$ and $r = r_1/r_2$.

9.5.5   Use of Scanning Electron-Beam Microscope

When the end surface of a fiber containing only one kind of dopant is chemically etched, the etching speed is a function of the dopant density. Therefore, after etching, a three-dimensional pattern appears on the fiber end. This pattern can be observed and recorded photographically by using an SEM (scanning electron-beam microscope). An example of the photograph is shown in Fig. 9.41 [28]. The refractive-index profile can be obtained from such a photograph in a somewhat qualitative manner.

Fig. 9.41   Electron-beam micrograph of a chemically etched fiber end (after Burrus *et al.* [28]).

9.6 Summary

Various methods for measuring the index profile of optical fibers and preform rods have been described. It is found that no method can satisfy the five requirements mentioned in Section 9.1. Proper choice must be made in actual measurement.

For multimode fibers, the microscopic interferometry using a sliced sample (method H in Table 9.1) is the most convenient and the most widely used. For single-mode fibers, this method is not suitable; instead, X-ray microanalysis (method K) and the scattering-pattern method (M) are usable. The former features high resolution, but the SN ratio is usually poor. The latter also features high resolution, but the drawback is the tedious data processing. Thus the advantages and importance of the index-profile measurement at preform stage is emphasized.

References

1. H. M. Presby, Refractive index and diameter measurement of unclad optical fibers, *J. Opt. Soc. Am.* **64**, No. 3, 280–284 (1974).
2. L. S. Watkins, Scattering from side-illuminated clad glass fibers for determination of fiber parameters, *J. Opt. Soc. Am.* **64**, No. 6, 767–772 (1974).
3. Y. S. Liu, Direct measurement of the refractive indices for a small numerical aperture cladded fiber: a simple method, *Appl. Opt.* **13**, No. 6, 1255–1256 (1974).
4. T. Kitano, H. Matsumura, M. Furukawa, and I. Kitano, Measurement of fourth-order aberration in a lens-like medium, *IEEE J. Quantum Electron.* **QE-9**, No. 10, 967–971 (1973).
5. Y. Suematsu and K. Furuya, A method for measuring the divergence of a beam spot in a lens-like medium and the group delay (in Japanese), Paper of Technical Group, IECE Japan, No. QE70-39 (December 1970).
6. E. G. Rawson and R. G. Murray, Interferometric measurement of SELFOC dielectric constant coefficient to sixth order, *IEEE J. Quantum Electron.* **QE-9**, No. 11, 1114–1118 (1973).
7. K. Maeda and J. Hamasaki, Measurement of fourth-order coefficient of refractive index of lens-like medium by using macroscopic intensity distribution of far-field pattern (in Japanese), Paper of Technical Group, IECE Japan, No. OQE75-73 (October 1975).
8. W. E. Martin, Refractive index profile measurement of diffused optical waveguides, *Appl. Opt.* **13**, No. 9, 2112–2116 (1974).
9. A. H. Cherin, L. Q. Cohen, W. S. Holden, C. A. Burrus, and P. Kaiser, Transmission characteristics of three Corning multimode optical fibers, *Appl. Opt.* **13**, No. 10, 2359–2364 (1974).
10. B. C. Wonsiewicz, W. G. French, P. D. Lasay, and J. R. Simpson, Automatic analysis of interferograms: optical waveguide refractive index profiles, *Appl. Opt.* **15**, No. 4, 1048–1052 (1976).
11. M. Ikeda, M. Tateda, and H. Yoshikiyo, Refractive index profile of a graded index fiber: measurement by a reflection method, *Appl. Opt.* **14**, No. 4, 814–815 (1975).

12. Y. Ueno and S. Aoki, The observation method of refractive index distribution of an optical fiber (in Japanese), Paper of Technical group, IECE Japan, No. OQE75-65 (September 1975).

13. W. Eickhoff and E. Weidel, Measuring method for refractive index profile of optical glass fibers, *Opt. Quantum Electron.* **7**, No. 2, 109–113 (1975).

14. S. Shiraishi and H. Takimoto, Spatial resolution in measuring refractive index profile of optical fibers by reflection method (in Japanese), Paper of Technical Group, IECE Japan, No. OQE75-97 (January 1976).

15. S. Aoki, S. Onoda, and M. Sumi, Measurement of refractive-index profile by vibrated reflection method (in Japanese), *Nat. Conv. Record, IECE Jpn.* **4**, 243, Paper No. 942 (March 1976).

16. D. Gloge and E. A. J. Marcatili, Multimode theory of graded-core fibers, *Bell Syst. Tech. J.* **52**, No. 9, 1563–1578 (1973).

17. M. J. Adams, D. N. Payne, and F. M. E. Sladen, Leaky rays on optical fibers of arbitrary (circularly symmetric) index profiles, *Electron. Lett.* **11**, No. 11, 238–240 (1975).

18. D. N. Payne, F. M. E. Sladen, and M. J. Adams, Index profile determination in graded index fibers, paper read at the *Eur. Conf. Opt. Fiber Commun., 1st*, pp. 43–45 (September 1975).

19. S. Tanaka, K. Inada, T. Akimoto, and T. Yamada, Measurement of refractive-index profile of optical fibers by near-field pattern method (in Japanese), *Nat. Conv. Record, IECE Jpn.* **4**, 236, Paper No. 935 (March 1976).

20. H. Osanai, O. Fukuda, Y. Takahashi, and M. Araki, Measurement of index profile of optical fibers by XMA (in Japanese), *Nat. Conv. Record, IECE Jpn.* **4**, 239, Paper No. 938 (March 1976).

21. S. Shiraishi, G. Tanaka, S. Suzuki, and S. Kurosaki, A simple method for measuring index profile of optical fibers (in Japanese), *Nat. Conv. Record, IECE Jpn.* **4**, 893, Paper No. 891 (March 1975).

22. A. M. Hunter II and P. W. Schreiber, Mach-Zehnder interferometer data reduction method for refractively inhomogeneous test objects, *Appl. Opt.* **14**, No. 3, 634–639 (1975).

23. K. Iga, Y. Kokubun, and N. Yamamoto, "Measurement of index distribution of focusing fibers by transverse differential interferometry (in Japanese), Record of *Nat. Symp. Light Radio Waves, IECE Jpn.* Paper No. S3-1 (November 1976).

24. T. Okoshi, K. Hotate, and H. Nomura, Computation of the refractive index distribution in an optical fiber from its scattering pattern for a normally incident laser beam (in Japanese), Paper of Technical Group, IECE Japan, No. OQE74-76 (February 1975).

25. T. Okoshi and K. Hotate, Automated measurement of refractive-index distribution in an optical fiber using the scattering pattern method (in Japanese), Paper of Technical Group, IECE Japan, No. OQE75-112 (February 1976).

26. T. Okoshi and K. Hotate, Computation of the refractive index distribution in an optical fibre from its scattering pattern for a normally incident laser beam, *Opt. Quantum Electron.* **8**, No. 1, 78–79 (1976).

27. T. Okoshi and K. Hotate, Refractive-index profile of an optical fiber: its measurement by the scattering-pattern method, *Appl. Opt.* **15**, No. 11, 2756–2764 (1976).

28. C. A. Burrus and R. D. Standley, Viewing refractive-index profiles and small-scale inhomogeneities in glass optical fibers: some techniques, *Appl. Opt.* **13**, No. 10, 2365–2369 (1974).

29. S. Chigira and R. Yamauchi, Measurement of structure of optical fiber-preform by back- and forward-scattered light (in Japanese), Paper of Technical Group, IECE Japan, No. OQE76-1 (April 1976).

30. T. Okoshi, K. Hotate, T. Yoshimura, and H. Shimoyama, Measurement of refractive index distribution of preform rods by interferometry I (in Japanese), Paper of Technical Group, IECE Japan, No. OQE76-90 (January 1977).

31. C. E. Pearson, "Handbook of Applied Mathematics," p.602. Van Nostrand Reinhold, New York, 1974.

32. J. L. Lundberg, Light scattering from large fibers at normal incidence, *Colloid Intface* **29**, No. 3, 565–583 (1969).

33. K. Hotate and T. Okoshi, Semiautomated measurement of refractive-index profiles of single-mode fibers by scattering pattern method, *Trans. IECE Jpn. Section E* **E61**, No. 3, 202–205 (1978).

34. K. Iga, Y. Kokubun, and N. Yamamoto, Refractive index profile measurements of focusing fibers by using a transverse differential interferometry (in Japanese), Paper of Technical Group, IECE Japan, No. OQE76-80 (December 1976).

35. G. D. Kahl and D. C. Mylin, Refractive deviation errors of interferograms, *J. Opt. Soc. Am.* **55**, No. 4, 364–372 (1965).

36. Y. Kokubun and K. Iga, Index profile measurement of optical fibers by means of a transverse interferogram—further considerations on its accuracy, *Trans. IECE Jpn. Sect. E* **E61**, No. 3, 184–187 (1978).

37. E. G. Rawson, D. R. Herriott, and J. McKenna, Analysis of refractive index distributions in cylindrical, graded-index glass rods (GRIN Rods) used as image relays, *Appl. Opt.* **9**, No. 3, 753–759 (1970).

38. K. Inada, Measurment of physical properties of optical fiber (in Japanese), *Joint Nat. Conv. Four Inst. Elec. Eng.* Paper No. 282 (October 1976).

10 | Measurement of Transmission Characteristics of Optical Fibers

Various methods for measuring the transmission characteristics of optical fibers are described. The methods can be classified into three groups: pulse method, swept-frequency modulation method, and spectrum analysis method. Typical examples of these methods are described, and methods for measuring multimode dispersion separately by using two (or more) light sources are presented. Two special measuring techniques are described: the shuttle-pulse method by which the measurement can be performed with relatively short samples, and the method for measuring the multimode dispersion of "mode groups" separately by using an optical mode filter.

10.1 Introduction

Group-delay characteristics of an optical fiber are important because they determine the maximum information–transmission rate of the communications channel using that fiber. Many of the previous chapters have discussed group-delay characteristics from a theoretical standpoint; this chapter is devoted to the description of their measurement.

Several basic concepts related to group-delay characteristics will be reviewed first.

10.1.1 Causes of Group-Delay Spread

There are three causes that bring about the spread of group delay (or dispersion*) in the transmission characteristics of an optical fiber:

* Strictly speaking, the term "group-delay characteristics" rather than "dispersion characteristics" should be used throughout, because the former expresses more directly the quantity actually measured. The reason that the latter is sometimes used in this chapter is that the terms "multimode dispersion," "waveguide dispersion," and "material dispersion" are now widely used to express pulse broadening.

236

(1)   multimode dispersion,
(2)   material dispersion, and
(3)   waveguide dispersion.

Detailed descriptions are given in Section 4.4.1. In multimode fibers, all three dispersions appear, whereas in single-mode fibers, the first one is absent, except for a similar but much smaller effect due to the presence of two orthogonal polarizations. This effect is called the polarization mode dispersion, or simply polarization dispersion.

10.1.2   Quantitative Expressions of Dispersions

In multimode dispersion, the complete quantitative description is given by the impulse response, or its Fourier transform which gives the complex frequency response of the dispersion. Typically, only one scalar quantity is used to express the dispersion; it is the maximum delay difference between modes [T; see Eq. (5.172)], or the rms deviation of the group delay [$Q(N)$; see Eq. (5.231)], or the 3-dB cutoff frequency of the frequency response [f_c; see Eq. (10.22)].

On the other hand, as described in Section 4.4.1, the pulse spread due to material and waveguide dispersions is proportional to the spectral (wavelength) bandwidth of the transmitted signal. Hence, these dispersions are sometimes called wavelength dispersions. The pulse spread due to wavelength dispersions is expressed as the product of the "specific dispersion" σ defined by Eq. (4.106) and the relative bandwidth of the transmitted signal $d\lambda/\lambda$ [see Eq. (4.105)]. When the specific dispersion σ is separated into the waveguide dispersion σ_w and the material dispersion σ_m,

$$\sigma_w = [\omega \, d^2\beta/d\omega^2]_{\text{waveguide}}, \tag{10.1}$$

$$\sigma_m = (\lambda^2/c) \, d^2 n_1/d\lambda^2. \tag{10.2}$$

For the magnitude of the actual impulse spread due to the preceding three factors, refer to Table 4.5.

10.2   Classification of Dispersion-Measurement Techniques

Methods of measuring dispersion characteristics are divided into three groups.

Pulse Method   Optical pulses are launched at the transmitting end, and the distorted output-pulse waveform is measured at the receiving end. The impulse response can be computed by "deconvoluting" the input-pulse waveform from the output one (see Sections 10.3.3–10.3.5).

Swept-Frequency Modulation Method   The output of a laser is amplitude-modulated by a swept-frequency sinusoidal signal, and is launched at the transmitting end. At the receiving end, the baseband-frequency response (the response in terms of the modulation frequency) is recorded.

Spectrum Analysis Method   The spectrum of a laser consists of a number of sidebands (e.g., with 100-MHz separation in a standard Kr–ion laser) due to the finite length of the resonator. Such an optical signal is transmitted through the fiber being tested, and the baseband-frequency response is obtained directly by comparing the input and output spectra.

On the other hand, methods of dispersion measurement can also be classified from another point of view; i.e., whether the separation of the multimode dispersion from the other two dispersions (the wavelength dispersions) is possible or not.

Total Dispersion Measurement   Those methods by which only the total dispersion can be measured will hereafter be called the total dispersion measurement. All methods using a single fixed-wavelength light source are included in this category.

Separable Dispersion Measurement   Those methods by which the multimode and wavelength dispersions can be measured separately will hereafter be called the separable dispersion measurement. The separable measurement is possible when a tunable laser, or two (or more) fixed-wavelength lasers, or a laser and a light-emitting diode (LED) are used.

The waveguide and material dispersions cannot be measured separately after the fiber is drawn. The only method for the separate measurement is to measure the material dispersion σ_m of the glass material beforehand at bulk state and subtract it from the total wavelength dispersion σ to obtain the waveguide dispersion σ_w. However, in most cases (in the 850-nm-wavelength region) $\sigma_w \ll \sigma_m$, and the value of σ_w can hardly be determined precisely.

Combining the two kinds of classification just described, we obtain six combinations as shown in Table 10.1. So far, experiments of 1-A, 1-B, 2-A, 2-B, and 3-A have been reported, but none on 3-B has appeared.

Somewhat special measurement techniques are the shuttle-pulse method by which the measurement can be performed with relatively short samples, and the method for measuring the multimode dispersion of mode groups separately by using an optical mode filter. These will be described later in the chapter.

Table 10.1

Classification of Methods of Dispersion Measurement

(1) Pulse method	
(2) Swept-frequency modulation method	(A) Total dispersion measurement
(3) Spectrum analysis method	(B) Separable dispersion measurement

10.3   Pulse Method

10.3.1   Measuring System

A typical measuring system is shown in Fig. 10.1 [1]. The components used in such a measuring system will be described, based on the state-of-the-art. The light source is a semiconductor laser in most cases. A typical GaAs–GaAlAs laser can deliver light pulses of 0.2 to 1.0 ns. The oscillation wavelength can be varied between 850 and 900 nm by controlling the composition of the material. The bandwidth of the oscillation spectrum (wavelength) is typically between 10^{-4} and 1.0 nm, primarily in accordance with the injection current level and the laser structure (see Table 4.5). In addition to the semiconductor laser, Nd–glass, Kr–ion, or Nd–YAG lasers are used.

The avalanche photo diodes used in detecting transmitted and received light pulses (shown as APD in Fig. 10.1) usually have a cutoff frequency of 1 to 3 GHz.

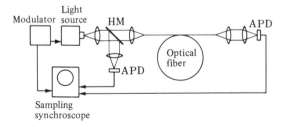

Fig. 10.1   A typical measuring system for the pulse method.

10.3.2   Impulse Response

Let the input-pulse (power) waveform, output-pulse waveform, and impulse response of an optical fiber be denoted by $x(t)$, $y(t)$, and $h(t)$, respectively. If $x(t) = 0$ for $t < 0$, the following convolution relation holds for these functions:

$$y(t) = \int_0^t h(t - \tau)x(\tau)\,d\tau = h(t) * x(t), \tag{10.3}$$

where the asterisk $(*)$ denotes a convolution integral. Equation (10.3) differs from ordinary linear convolution in that $x(t)$, $y(t)$, and hence $h(t)$ must always be positive, because $x(t)$ and $y(t)$ represent power waveforms.

It is known that the foregoing linear response for the power waveform holds when the following two conditions are satisfied [2].

(1)   The spectral spread due to the pulse modulation is negligible compared with the original spectral spread of the light source.

(2)   The oscillation spectrum does not fluctuate appreciably during the pulse duration.

In most cases, condition (1) is satisfied as shown in Table 4.5 and Fig. 4.12b. As to condition (2), an external pulse modulator should be used if very precise measurement is required.

The impulse response $h(t)$ is dependent on the power distribution among modes at the launch end. Therefore, in the measurement of $h(t)$, a mode scrambler (a pseudorandomly curved short fiber section) is often inserted at the launch end to standardize the power distribution. Nevertheless, when we measure the impulse response of a "cascaded" fiber, it is not always equal to the convolution of those of two constituent fibers. It is because the power distribution at the connection (i.e., the input of the second fiber) becomes more or less different from the standardized one due to the mode-conversion effect. The effect of the mode conversion is usually negligible when the fiber is shorter than several hundred meters (see Fig. 8.1). Except for those fibers in which a structural nonuniformity is provided along the fiber axis intentionally to enhance intermodal coupling, the critical fiber length (inflection point in Fig. 8.1) is apt to become longer year by year as the result of the progress in manufacturing technology [3].

10.3.3   Simple Expression for Impulse Response

A convenient measure of the quality of the impulse response $h(t)$ is its spread width. We assume for simplicity that $x(t)$, $y(t)$, and $h(t)$ all have Gaussian waveforms with widths $\Delta\tau_1$, $\Delta\tau_2$, and $\Delta\tau$, respectively. Under such an assumption, we may write

$$x(t) = \exp[-(2t/\Delta\tau_1)^2], \tag{10.4}$$

$$y(t) = \exp[-\{2(t - t_a)/\Delta\tau_2\}^2], \tag{10.5}$$

$$h(t) = \exp[-\{2(t - t_a)/\Delta\tau\}^2], \tag{10.6}$$

where t_a is the average delay time. Substituting these equations into

$$y(t) = h(t) * x(t), \tag{10.7}$$

we obtain

$$\Delta\tau_2^2 = \Delta\tau_1^2 + \Delta\tau^2. \tag{10.8}$$

In many cases, the spread of the impulse response $\Delta\tau$ is computed from measured values of $\Delta\tau_1$ and $\Delta\tau_2$ using Eq. (10.8). A similar assumption is also made to separate the multimode and wavelength dispersions (see Section 10.3.6).

10.3.4 Exact Computation of Impulse Response—Fourier-Transform Method

When the waveforms of $x(t)$ and $y(t)$ are given, a standard method for computing $h(t)$ is the Fourier-transform method. We denote the Fourier transforms of $y(t)$, $h(t)$, and $x(t)$ by $Y(\omega)$, $H(\omega)$, and $X(\omega)$, respectively. $Y(\omega)$ is expressed as

$$Y(\omega) = H(\omega)X(\omega). \tag{10.9}$$

Therefore, expressing the inverse Fourier transform by $\mathcal{F}^{-1}$, we may write

$$h(t) = \mathcal{F}^{-1}[H(\omega)] = \mathcal{F}^{-1}[Y(\omega)/X(\omega)]. \tag{10.10}$$

Computation of $H(\omega)$ and $h(t)$ based on Eq. (10.10) is common practice (e.g., see Horiguchi *et al.* [4] for the computation of $H(\omega)$, and Timmermann *et al.* [2] for $h(t)$.) Furthermore, if the smoothed waveforms are obtained from the sampling synchroscope shown in Fig. 10.1 and fed into a computer in which the FFT (fast Fourier transform) computation can be readily performed, the function $h(t)$ can be displayed on a CRT on a real-time basis.

10.3.5 Exact Computation of Impulse Response—Direct Deconvolution Method

An alternative, more direct method for computing the impulse response is the direct deconvolution method [1]. We first note the temporal relation between $x(t)$, $h(t)$, and $y(t)$ shown in Fig. 10.2; the average delay time is much longer than the pulse widths. We assume that $x(t)$ and $y(t)$ start to rise at $t = 0$ and $t = t_0$, respectively. Then $h(t)$ also starts to rise at $t = t_0$.

Next, we divide each function by time increments Δt, and use the central values in each increment for $x(t)$ and $h(t)$, and the final values in each increment for $y(t)$ as their sampled values. Then Eq. (10.3) is rewritten in an

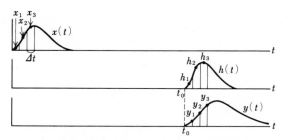

Fig. 10.2   The relation between $x(t)$, $h(t)$, and $y(t)$.

incremental form

$$y(t) = 0 \qquad (0 \leqq t < t_0), \tag{10.11a}$$

$$y_1 = y(t_0 + \Delta t) = \int_0^{t_0 + \Delta t} h(t_0 + \Delta t - \tau)x(\tau)\, d\tau$$

$$= \int_0^{\Delta t} h(t_0 + \Delta t - \tau)x(\tau)\, d\tau$$

$$= h_1 x_1 \Delta t, \tag{10.11b}$$

$$y_2 = y(t_0 + 2\Delta t) = (h_2 x_1 + h_1 x_2)\Delta x, \tag{10.11c}$$

$$\vdots$$

$$y_n = y(t_0 + n\Delta t) = \left(\sum_{k=1}^{n} h_{n-k+1} x_k \right)\Delta x, \tag{10.11d}$$

which can be expressed in a unified form

$$\underbrace{\begin{bmatrix} y_1 \\ y_2 \\ \vdots \\ y_n \end{bmatrix}}_{\mathbf{Y}_0} = \underbrace{\begin{bmatrix} x_1 & 0 & \cdots & \cdots & 0 \\ x_2 & & & & \vdots \\ & & & & \\ \vdots & & & & 0 \\ x_n & \cdots & \cdots & x_2 & x_1 \end{bmatrix}}_{\mathbf{X}_0} \underbrace{\begin{bmatrix} h_1 \\ h_2 \\ \vdots \\ h_n \end{bmatrix}}_{\mathbf{H}_0} \Delta t. \tag{10.12}$$

Therefore, if the inverse of $\mathbf{X}_0$ exists, denoted by $\mathbf{X}_0^{-1}$, the impulse response $\mathbf{H}_0$ can be obtained as

$$\mathbf{H}_0 = (1/\Delta t)\mathbf{X}_0^{-1}\mathbf{Y}_0. \tag{10.13}$$

Actually, however, the impulse response $\mathbf{H}_0$ thus obtained diverges. To prevent this, we assume that $h_k = 0$ for $k \geq n + 1$, where n is an appropriate integer. Under such an assumption, Eq. (10.12) becomes

$$
\begin{bmatrix} y_1 \\ \vdots \\ y_n \\ \vdots \\ y_l \end{bmatrix}_{\mathbf{Y}} = \begin{bmatrix} x_1 & 0 & \cdots\cdots & 0 \\ \vdots & & & \vdots \\ & & & 0 \\ x_n & \cdots\cdots & & x_1 \\ \vdots & & & \vdots \\ x_l & \cdots\cdots & & x_{l-n+1} \end{bmatrix}_{\mathbf{X}} \begin{bmatrix} h_1 \\ \vdots \\ h_n \end{bmatrix}_{\mathbf{H}} \Delta t, \qquad (10.14)
$$

where $l > n$. This gives a set of linear simultaneous equations whose number is greater than the number of variables. Therefore, no exact solution can exist; an approximate solution follows the least squares method. We rewrite Eq. (10.14) using $\mathbf{Y}_r = \mathbf{Y}/\Delta t$ and determine $\mathbf{H}_{opt}$ with which

$$P = (\mathbf{Y}_r - \mathbf{X}\mathbf{H})^t(\mathbf{Y}_r - \mathbf{X}\mathbf{H}) \qquad (10.15)$$

is minimized, where the superscript t denotes a matrix transposition. From the condition $dP/d\mathbf{H} = 0$, $\mathbf{H}_{opt}$ is obtained as

$$\mathbf{H}_{opt} = (\mathbf{X}^t\mathbf{X})^{-1}\mathbf{X}^t\mathbf{Y}(1/\Delta t). \qquad (10.16)$$

Described in the foregoing is the general method for computing an impulse response. However, in the measurement of the transmission characteristics of an optical fiber, $\mathbf{H}_{opt}$ thus obtained consists of widely scattered points. The following smoothing technique (which is a standard technique in the field of system identification) should be employed to reduce scattering [5].

Figure 10.3 shows an example of the scattered impulse response computed by using Eq. (10.16). Generally, by integrating an impulse response with

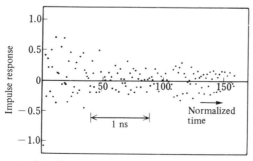

Fig. 10.3   An example of the "scattered" impulse response obtained by using Eq. (10.16).

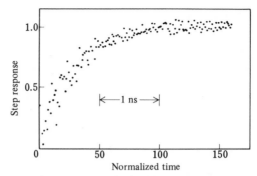

Fig. 10.4   The scattered step response obtained by integrating the impulse response shown in Fig. 10.3.

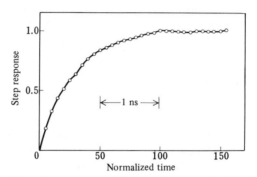

Fig. 10.5   The step response obtained by smoothing Fig. 10.4.

respect to time, we obtain the step response, and by differentiating the latter, we obtain the former. If we integrate the data in Fig. 10.3, which looks hopeless, we obtain the step response shown in Fig. 10.4. Note that the ordinate of the step response shown in Fig. 10.4 is normalized to unity.

It is not difficult to smooth out scattering in the step response shown in Fig. 10.4; e.g., we can average the data locally to obtain a smoother response. The result of such smoothing is shown in Fig. 10.5. The impulse response can then be computed by differentiating Fig. 10.5; the result is shown in Fig. 10.6.

To prove the validity of the impulse response thus obtained, we synthesize the output waveform by convoluting the input waveform and the computed impulse response, and compare the result with the measured output waveform. Figure 10.7 shows an example of such comparison; the solid and dashed curves show the measured and synthesized output waveforms, respec-

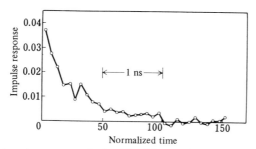

Fig. 10.6   The impulse response obtained by differentiating the response of Fig. 10.5.

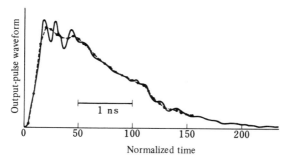

Fig. 10.7   Measured output waveform (solid curve) and synthesized output waveform (dashed curve) obtained by convoluting the input waveform and the computed impulse response (after Okoshi and Sasaki [1]).

tively [1]. The overall shapes of the waveforms show good agreement, but small ripples are missing in the synthesized waveform.

10.3.6   Separable Dispersion Measurement

Several papers [6, 7] have reported the separable dispersion measurement, i.e., the measurement by which multimode and wavelength dispersions are obtained separately. The measuring system and the results described in Ref. [7] will be described here as an example.

As shown in Fig. 10.8, outputs of two GaAs lasers oscillating at $\lambda_1 = 860$ nm and $\lambda_2 = 900$ nm are launched on the fiber being tested (about 1 km long) through a half mirror. Figures 10.9a and 10.9b show the measured input and output pulse waveforms. In both figures, the left and right pulses show those at λ_2 (900 nm) and λ_1 (860 nm), respectively.

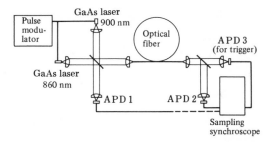

Fig. 10.8 A typical measuring system using two lasers for the seprarable dispersion measurement (after Gloge *et al.* [7]).

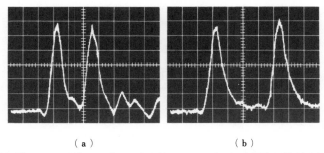

(a)   (b)

Fig. 10.9 The measured waveforms in the system shown in Fig. 10.8: (a) transmitted pulses and (b) received pulses. In both figures, the left and right pulses correspond to the λ_2 and λ_1 pulses, respectively, where $\lambda_2 = 900$ nm and $\lambda_1 = 860$ nm (after Gloge *et al.* [7]).

The principle of the separable measurement is as follows. The difference in the delay time between the λ_1 and λ_2 pulses is produced solely by the waveguide and material dispersions, independently of the multimode dispersion. Hence, the difference in the flight time between the λ_1 and λ_2 pulses can be expressed as

$$\Delta t_{\mathrm{a}} = (\sigma_{\mathrm{w}} + \sigma_{\mathrm{m}})(\lambda_2 - \lambda_1)/\lambda, \tag{10.17}$$

where $\lambda = (\lambda_1 + \lambda_2)/2$. Since Δt_{a} can be readily obtained from measured waveforms such as Figs. 10.9a and 10.9b, we can compute $(\sigma_{\mathrm{w}} + \sigma_{\mathrm{m}})$ using Eq. (10.17).

Next, we compare the pulse widths in Figs. 10.9a and 10.9b. We assume for simplicity that the transmitted pulse and the impulse responses due to multimode dispersion, waveguide dispersion, and material dispersion all have Gaussian waveforms; we denote their widths by $\Delta\tau_1$, $\Delta\tau_{\mathrm{M}}$, $\Delta\tau_{\mathrm{w}}$, and

$\Delta\tau_m$, respectively. Then the width of the received pulse is

$$\Delta\tau_2^2 = \Delta\tau_1^2 + \Delta\tau_M^2 + (\Delta\tau_w + \Delta\tau_m)^2. \tag{10.18}$$

On the other hand, if the relative bandwidth of the laser output is denoted by $(\Delta\lambda/\lambda)$, then

$$\Delta\tau_w + \Delta\tau_m = (\Delta\lambda/\lambda)(\sigma_w + \sigma_m). \tag{10.19}$$

We can compute the pulse spread due to multimode dispersion $\Delta\tau_M$ using Eqs. (10.18) and (10.19).

In actual measurements, $\Delta\tau_M$ cannot always be computed precisely because sometimes $\Delta\tau_1 \simeq \Delta\tau_2$, as is the case in Fig. 10.9. In Gloge *et al.* [7], the pulse widths are determined by a separate, more precise measurement.

10.4   Swept-Frequency Modulation Method

10.4.1   Measuring System

In the swept-frequency modulation method, an optical setup identical to that shown in Fig. 10.1 is used, but the modulation circuit is different in that the light intensity is modulated by a swept-frequency sinusoidal signal. At the receiving end, the light is detected by the APD, and the detected signal is amplified and fed to a voltmeter to obtain directly the response of the fiber with respect to the modulation frequency, i.e., $H(\omega)$ in Eq. (10.9).

A comment should be added concerning a practically useful relation between the impulse and frequency responses. If we assume a Gaussian impulse response as shown in Eq. (10.6), from a Fourier-transform formula,* the frequency response can be expressed as

$$|H(\omega)| = |\mathscr{F}[h(t)]| = \mathscr{F}[\exp\{-(2t/\Delta\tau)^2\}] \propto \exp\{-(\omega\,\Delta\tau/4)^2\}. \tag{10.20}$$

On the other hand, if we denote the frequency at which $H(\omega)$ is halved $(-3\ \text{dB})$ by f_c, we may write

$$|H(\omega)| \propto \exp\{-(\log_e 2)(\omega/2\pi f_c)^2\}. \tag{10.21}$$

---

* If we define the Fourier transform of a function $f(t)$ as

$$F(\omega) = \mathscr{F}[f(t)] = \frac{1}{2\pi}\int_{-\infty}^{\infty} e^{j\omega t} f(t)\, dt,$$

then we obtain $\mathscr{F}[\exp(-at^2)] = \exp(-\omega^2/4a)/\sqrt{2a}$.

Comparing Eqs. (10.20) and (10.21), we obtain

$$f_c = (2\sqrt{\log_e 2}/\pi)(1/\Delta\tau) = 0.530(1/\Delta\tau). \qquad (10.22)$$

This is the relation to be obtained. For example, when $\Delta\tau = 1$ ns, the cutoff frequency $f_c = 530$ MHz.

10.4.2   Example of Total Dispersion Measurement

At present, the swept-frequency-modulation, total-dispersion measurement is the most commonly performed; it is rather difficult to choose an example from many reports. One of the earliest reports by Personick [8] will be described.

The circles in Fig. 10.10 show the result of the measurement of the baseband-frequency response $H(\omega)$ of a 1-km fiber, by using a GaAlAs LED modulated by a swept-frequency signal as the light source. The solid curve is the result of curve fitting in which the impulse response is assumed to be Gaussian, and from which f_c is determined by the least squares method $(1/2\pi f_c = 1.75$ ns). Computation of the impulse-response width from this f_c value leads to $\Delta\tau = 4.1$ ns, which is in good agreement with $\Delta\tau = 3.8$ nm obtained directly by the pulse method [8].

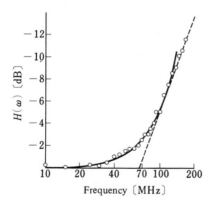

Fig. 10.10   A measured baseband frequency response of a 1-km fiber (after Personick [8]).

10.4.3   Example of Separable Dispersion Measurement

Kobayashi *et al.* [9] proposed a swept-frequency-modulation, separable-dispersion-measurement scheme. In their method, two laser lights having different wavelengths are modulated by a common sinusoidal signal and launched on the fiber, and the wavelength dispersion is measured separately from the phase relation between the demodulated signals at the receiving end.

Let the two wavelengths be denoted by λ_1 and λ_2, respectively, their average by $\lambda \, [= (\lambda_1 + \lambda_2)/2]$, and their difference by $\Delta\lambda \, (= \lambda_2 - \lambda_1 \ll \lambda)$. When the fiber length is L, the delay-time difference between the λ_1 and λ_2 lights is expressed as

$$\Delta t_a = (L \, \Delta\lambda/\lambda)(\sigma_w + \sigma_m). \tag{10.23}$$

At the receiving end, the two lights are detected simultaneously by a common APD, and the demodulated signal level is recorded as a function of the modulation frequency f_m. When

$$\Delta t_a = (1/2f_m)(2n + 1) \qquad (n = \text{integer}), \tag{10.24}$$

the two signal components cancel, and the demodulated signal level is minimized. Hence, if we measure two modulation frequencies which give the minimized signal level, we can compute Δt_a from the difference between those frequencies, and compute $(\sigma_w + \sigma_m)$ by using Eq. (10.23). The multimode dispersion can be obtained by subtracting the wavelength dispersion thus obtained from the total dispersion measured by the standard method.

The principal advantage of this method is that a precise measurement is possible, because (1) this is a kind of null method, and (2) in the present state of the art, the frequency measurement can offer the highest accuracy among measurements of various physical quantities.

10.5   Spectrum Analysis Method

Spectrum analysis is not a widely used method. The experiment by Gloge *et al.* [10] will be described.

A free-running Kr-ion laser is used as the light source. As shown in Fig. 10.11a, the output power spectrum of a Kr-ion laser consists of a number of sidebands whose separation is determined by the resonator length; in the present case the separation is about 100 MHz. Figure 10.11b shows the light spectrum after it travels through a 30-m-long fiber.

The baseband-frequency response $H(\omega)$ can be readily obtained by comparing Figs. 10.11a and 10.11b. The result is shown in Fig. 10.12 by dots; the solid curve shows the frequency response computed by assuming that the impulse response $h(t)$ is a square function with width $T = 0.6$ ns. Under such an assumption, the frequency response is

$$H(\omega) = \sin(T\omega/2)/(T\omega/2). \tag{10.25}$$

Fairly good agreement is found between the solid curve and the dots.

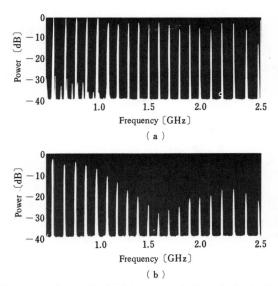

Fig. 10.11   The spectra observed in the spectrum analysis method: (a) transmitted spectrum and (b) received spectrum (after Gloge *et al.* [10]).

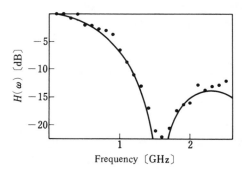

Fig. 10.12   The baseband frequency response obtained by comparing Figs. 10.11a and 10.11b (after Gloge *et al.* [10]).

10.6   Shuttle-Pulse Method

As transmission loss is reduced, we will be concerned with the transmission characteristics of long fibers, e.g., fibers longer than 10 km. However, such long fibers are inconvenient for measurement; moreover, they must be cut into shorter pieces when transmission characteristics are to be measured as functions of fiber length. To overcome these difficulties, the shuttle-pulse

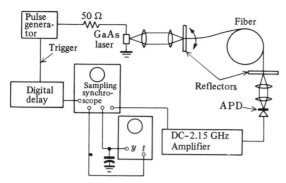

Fig. 10.13   An example of the measuring system for shuttle-pulse method (after Cohen [11]).

method originally developed in the millimeter-wave region is now becoming prevalent.

An example of the measuring system is shown in Fig. 10.13 [11]. Semi-transparent reflectors mounted on adjustable pedestals are placed at both ends of a relatively short sample (160 m in the present case) of the fiber to be tested. The launched light pulse is shuttled between the two semitransparent reflectors (half mirrors), but every time the pulse arrives at the receiving end, a part of the pulse energy is picked up through the half mirror and the pulse waveform is observed. The reflector consists of a precisely bored cylinder and a half mirror fixed normal to it as shown in Fig. 10.14. The gap between the half mirror and the fiber end is filled with matching oil.

Fig. 10.14   The reflecting device used in the shuttle-pulse method (after Cohen [11]).

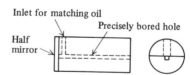

Figure 10.15a shows the entire waveform of pulses (from the first through the fifth pulses), and Figs. 10.15b–10.15d show the first, fifth, and tenth pulses in an enlarged time scale, respectively. Unless the reflecting device causes additional conversion between propagating modes, the waveforms in Figs. 10.15b–10.15d should show those at propagation distances of 106, 954, and 2014 m, respectively.

The optimum design of the measuring system will now be considered. The most important design parameter in such a measuring system is the power reflection coefficient R of the half mirror. When R is too large, the transmission coefficient $T (= 1 - R)$ becomes excessively small, thereby reducing the output light power and making precise measurement difficult.

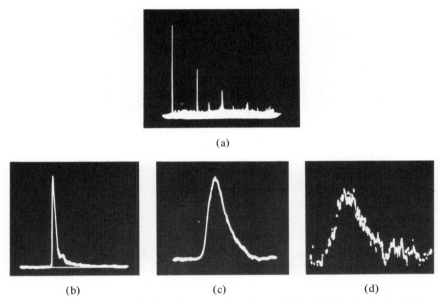

Fig. 10.15 Waveforms obtained in the shuttle-pulse method: (a) entire waveform of pulses, (b) enlarged first pulse ($L = 106$ m), (c) enlarged fifth pulse ($L = 945$ m), (d) enlarged tenth pulse ($L = 2014$ m) (after Cohen [11]).

On the contrary, when R is too small, the propagating power is attenuated drastically due to loss at both ends, and the long-distance characteristics cannot be measured.

We denote the input pulse and the Nth output pulse powers by P_{in} and $P_{out}(N)$, respectively. If we assume first that the fiber is lossless, then from the schematic representation in Fig. 10.16, the Nth pulse output power is

$$P_{out}(N) = P_{in}T^2R^{2(N-1)} = P_{in}(1 - R)^2R^{2(N-1)}. \qquad (10.26)$$

In actual fibers, the light pulse is subject to attenuation, so that Eq. (10.26) is modified to

$$P_{out}(N) = P_{in}(1 - R)^2R^{2(N-1)}\exp\{-\alpha L(2N - 1)\}, \qquad (10.27)$$

where α is the attenuation constant or in decibel representation,

$$P_{out}(N)\,[\text{dB}] = P_{in}[\text{dB}] + \{20\log(1 - R)R^{N-1} - 4.34L(2N - 1)\alpha\}\,[\text{dB}].$$
$$(10.28)$$

Fig. 10.16   Schematic representation of the Nth pulse in the shuttle-pulse method.

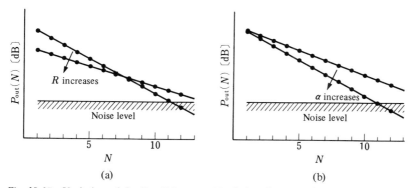

Fig. 10.17 Variation of the P_{out} (N)-versus-N relation due to variations of (a) parameter R, and (b) parameter α.

Figures 10.17a and 10.17b show the variation of the $P_{out}(N)$-versus-N relation with respect to variations of R and α, respectively. These figures show that the reflection coefficient R should be determined by a compromise taking into account the attenuation constant α, the threshold noise level of the amplifier, and the equivalent fiber length $(2N - 1)L$ over which the measurement is to be performed. Generally speaking, when $(2N - 1)L$ is large and/or α is small, R should be large.

In Cohen [11], measurement of the mode mixing using the shuttle-pulse method is described. In Ikeda *et al.* [12], the "circulating-pulse method" is proposed, in which the optical pulse circulates in a ring-formed, relatively short fiber, while a part of the pulse power is taken out every time the pulse passes a branching point.

10.7 Measurement of Multimode Dispersion for Each Mode Group

The first purpose of the dispersion measurement is to know the characteristics of a given fiber for the design of the communication system using that fiber. The second purpose is to make use of the obtained data for the improvement of design and fabrication techniques. Technically the second purpose is more significant.

To make the multimode-dispersion measurement more useful for fiber design, Cohen proposed a simple method by which the multimode dispersion can be measured separately for each mode group [13]. Such a measurement can offer the desired information for multimode dispersion improvement.

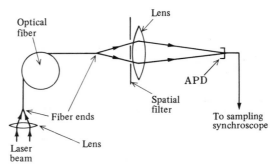

Fig. 10.18 Optical setup for measuring multimode dispersion for each mode group (after Cohen [13]).

The essential part of the optical setup is shown in Fig. 10.18. At the receiving end, the light bundle, before entering the detector, passes a simple spatial filter which consists of a convex lens and an appropriately designed annular (or circular) aperture. Generally, the light bundle from lower propagation modes passes the central part of the lens, whereas the light bundle from higher modes passes the peripheral part of the lens. Hence, what is detected by the APD is the pulse waveform for higher-order modes when the aperture is annular (as shown in Fig. 10.18), and is the pulse waveform for lower-order modes when the aperture is a circular one placed at the center of the lens. (If desired, we can divide the light energy into more than two groups by preparing many apertures.)

Figure 10.19 shows the output-pulse waveforms of α-power nonuniform-core fibers. The left and right columns show the waveforms for $\alpha \simeq 1.5$ and $\alpha > 2$, respectively. The top two waveforms show the total power waveforms, the middle two waveforms show those for lower-order modes ($0 < NA < 0.038$, where NA denotes numerical aperture; see Section 3.2.3), and the bottom two show those for higher-order modes ($0.112 < NA < 0.156$). In the left-hand figures ($\alpha = 1.5$), the higher modes arrive earlier than the lower modes, and in the right-hand figures ($\alpha > 2$), the order of their arrival is reversed.

Figure 10.20 shows the results of simultaneous measurement of the higher and lower modes for various α-power profiles. In Figs. 10.20a–10.20f, the values of the power α are 1.57, 1.63, 1.77, 1.87, 2.05, and 2.19, respectively. Other measurements can confirm that the energy transmitted by higher-order modes appears as a small forerunning pulse in Figs. 10.20a and 10.20b, and as a small following pulse in Figs. 10.20e and 10.20f. Therefore, these waveforms suggest that the optimum condition is given approximately by Fig. 10.20c, i.e., $\alpha = 1.77$.

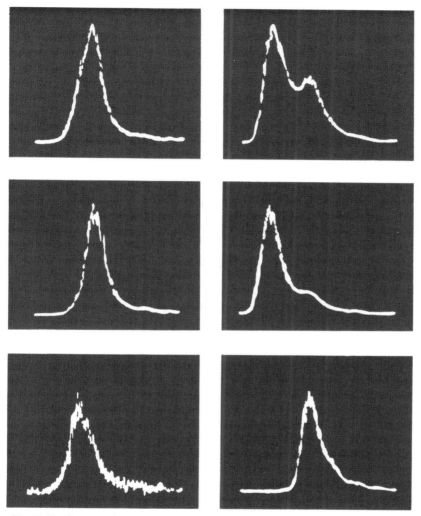

Fig. 10.19 Output-pulse waveforms of α-power nonuniform-core fibers obtained by the setup shown in Fig. 10.18 (after Cohen [13]). For details, refer to the text.

As shown in Eq. (5.166) or Section 5.7, the optimum power α is expressed, by using Olshansky's parameter y, as

$$\alpha_{\text{opt}} = 2 + y. \tag{10.29}$$

Since the y value of the material used in the fiber being tested is about -0.23, the foregoing optimum α agrees well with the theoretical prediction.

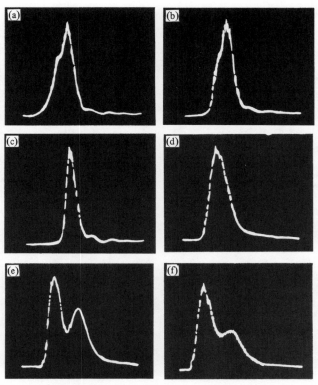

*Fig. 10.*20   Multimode dispersion for various α-power nonuniform-core fibers (after Cohen [13]). For details, refer to the text.

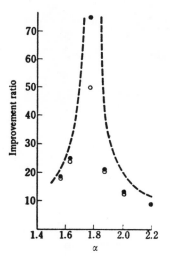

Fig. 10.21 The computed "improvement ratio" (dashed curves) and the measured improvement ratio (circles) (after Cohen [13]). The dots are explained in the text.

The dashed curves in Fig. 10.21 show the "improvement ratio" (multimode dispersion of a uniform-core fiber/multimode dispersion of an α-power fiber) computed by the method described in Section 5.4.3. The circles show the measured total dispersion, and the dots show the multimode dispersion obtained by subtracting the material dispersion (0.18 ns/km) from the measured total dispersion.

10.8 Summary

Various techniques for measuring the transmission characteristics of optical fibers have been described. Multimode fibers have primarily been considered. Nevertheless, most of the descriptions remain valid for single-mode fibers; however, technologically the measurement becomes more difficult because the dispersion becomes very small. For single-mode fibers, double-laser schemes are commonly used to expand the wavelength dispersion and facilitate its measurement [14].

References

1. T. Okoshi and K. Sasaki, Precise measurement of the impulse response of an optical fiber, *Trans. IECE Jpn. Sect. E*, **E61**, No. 12, 964–965 (1978).
2. C. C. Timmermann *et al.*, The experimental determination of the baseband transfer function of optical multimode fibers, *Arch. Elektronik Übertragungstechnik* **29**, Heft 5, 235–237 (1975).
3. E. L. Chinnock, L. G. Cohen, W. S. Holden, R. D. Standley, and D. B. Keck, The length dependence of pulse spread in the CGW-Bell-10 optical fiber, *Proc. IEEE* **61**, No. 10, 1499–1500 (1973).
4. M. Horiguchi, T. Edahiro, and T. Miyashita, Pulse transmission characteristics of multimode optical fibers (in Japanese), National Convention Record, IECE Japan, Paper No. 1161 (November 1974).
5. A. P. Sage, J. L. Melsa, "System Identification," pp. 6–13. Academic Press, New York, 1971.
6. D. Gloge, E. L. Chinnock, R. D. Standley, and W. S. Holden, Dispersion in a low-loss multimode fiber measured at three wavelengths, *Electron. Lett.* **8**, No. 21, 527–529 (1972).
7. D. Gloge, E. L. Chinnock, and T. P. Lee, GaAs twin-laser setup to measure mode and material dispersion in optical fibers, *Appl. Opt.* **13**, No. 2, 261–263 (1974).
8. S. D. Personick, W. M. Hubbard, and W. S. Holden, Measurements of the baseband frequency response of a 1-km fiber, *Appl. Opt.* **13**, No. 2, 266–268 (1974).
9. I. Kobayashi, K. Aoyama, and M. Koyama, Measurement of material dispersion of low-loss multimode optical fibers (in Japanese), *Record Nat. Symp. Light Radio Waves*, IECE *Jpn.* Paper No. S3-3 (November 1976).
10. D. Gloge, E. L. Chinnock, and D. H. Ring, Direct measurement of the (baseband) frequency response of multimode fibers, *Appl. Opt.* **11**, No. 7, 1534–1538 (1972).

11.   L. G. Cohen, Shuttle pulse measurements of pulse spreading in an optical fiber, *Appl. Opt.* **14**, No. 6, 1351–1356 (1975).
12.   M. Ikeda, T. Tanifuji, and Y. Yamauchi, Measurement of transmission characteristics of optical fibers by circulating pulse method (in Japanese), *Record Nat. Conv. Light Radio Waves, IECE Jpn.* Paper No. S3-12 (November 1976).
13.   L. G. Cohen, Pulse transmission measurements for determining near optimal profile gratings in multimode borosilicate optical fibers, *Appl. Opt.* **15**, No. 7, 1808–1814 (1976).
14.   J. Sakai, H. Tsuchiya, and T. Itoh, Transmission characteristics of single-mode fibers (in Japanese), National Convention Record, IECE Japan, Paper No. 928 (March 1975).

11 | Concluding Remarks

In this final short chapter, the important technical tasks at present and in the near future are mentioned.

11.1 Development of Practical Single-Mode Fibers

At present, the minimum transmission loss of a single-mode fiber has been reduced to about 0.5 dB/km. This value suggests that in the somewhat distant future, large-capacity, nationwide public communications systems will be constructed by using ultralow-loss, ultralow-dispersion single-mode fibers. Further loss reduction and cancellation of material and waveguide dispersions are required for this purpose.

11.2 Development of Long-Wavelength (1.0–1.6 μm) Optical-Fiber Communications System

In recent low-loss silica fibers, the lowest transmission loss is usually obtained at wavelengths a little longer than standard GaAlAs-laser wavelengths, i.e., in the 1.0–1.6 μm range (see Fig. 2.6). Besides, in the vicinity of 1.3 μm, the material and waveguide dispersions cancel each other without any elaborate profile shaping (see Section 7.2.1). Therefore, the developments of the optical fibers as well as optoelectronic devices (lasers and photodetectors) for use in this wavelength range are important technical tasks.

259

11.3   Development of FDM (Frequency-Division Multiplexing) Optical-Fiber Communications System

To make effective use of the wide bandwidth available in an optical fiber (note that the wavelength range 1.0–1.6 μm corresponds to a frequency bandwidth of 120,000 GHz), the FDM technique will be developed in the future. In an FDM system, many optical signals are combined together at the transmitting end by using a multiplexer and are divided into the original number of channels at the receiving end by using a demultiplexer (branching filter). The development of multiplexers and demultiplexers, as well as of a series of lasers covering the low-loss frequency range of the fiber, will become significant tasks.

11.4   Development of Low-Cost, Large-Diameter, Large-Numerical-Aperture Fibers

Along with the high-performance fibers considered in the preceding three sections, low-cost, easy-to-handle fibers for use in telephone subscriber lines, cable television systems, computer networks, airplanes, and vehicles will also be developed. In such fibers, a large diameter and a large numerical aperture might be required to facilitate the fiber-to-fiber splicing/connection and laser-to-fiber coupling.

11.5   Development of Economical Methods for Manufacturing Optical Fibers

At present, high-performance fibers are very expensive. To widen the applications of optical fibers in the future, more economical methods for manufacturing fibers must be devised and developed.

There are other important tasks, such as the development and refinement of coupling/splicing techniques, and standardization. However, optical fiber communications is an inherently "gifted" technique. Despite many technical problems to be encountered in future, the optical fiber will finally reach the major position among various information–transmission schemes in human society.

Appendix

2A.1 Derivation of Eqs. (2.68) and (2.69)

From Eq. (2.64),

$$(\nabla\phi)^2 = n^2. \tag{2A.1}$$

Taking gradients of both sides, we obtain

$$2\nabla\phi\,\nabla^2\phi = 2n\,\nabla n. \tag{2A.2}$$

This equation and Eq. (2.65) lead to

$$\mathbf{S}\nabla^2\phi = \nabla n. \tag{2A.3}$$

On the other hand, generally

$$\frac{d}{ds} = \sum_{i=1}^{3} \frac{dx_i}{ds}\frac{\partial}{\partial x_i} = \frac{d\mathbf{r}}{ds}\cdot\nabla \qquad (x_i = x,\,y,\,z) \tag{2A.4}$$

Using Eqs. (2.65), (2A.4), (2.66), and (2A.3), we obtain

$$\frac{d}{ds}\left[n\mathbf{s}\right] = \frac{d}{ds}\nabla\phi = \frac{d\mathbf{r}}{ds}\nabla^2\phi = \mathbf{s}\nabla^2\phi = \nabla n, \tag{2A.5}$$

which is Eq. (2.68). Equation (2.69) can be derived immediately from Eq. (2.68).

3A.1 Derivation of Eq. (3.16)

We note first that

$$L_1 = L_0/n_1, \qquad M_1 = M_0/n_1. \tag{3A.1}$$

On the other hand, for path s_1 shown in Fig. 3.3,

$$y - y_0 = (M_1/L_1)(x - x_0); \tag{3A.2}$$

hence,

$$y = (M_1/L_1)x + [y_0 - (M_1/L_1)x_0]. \tag{3A.3}$$

For the first reflection point (x_1, y_1, z_1), $x_1^2 + y_1^2 = a^2$. Therefore, at this point, from Eq. (3A.3),

$$\left(1 + \frac{M_1^2}{L_1^2}\right)x_1^2 + 2\frac{M_1}{L_1}\left(y_0 - \frac{M_1}{L_1}x_0\right)x + \left(y_0 - \frac{M_1}{L_1}x_0\right)^2 - a^2 = 0,$$

and hence

$$x_1 = \frac{-(L_1 M_1 y_0 - M_1^2 x_0) \pm L_1[(L_1^2 + M_1^2)a^2 - (L_1 y_0 - M_1 x_0)^2]^{1/2}}{L_1^2 + M_1^2} \tag{3A.4}$$

We next compute the left-hand side of Eq. (3.16). Using Eqs. (3A.3) and (3A.4), we obtain

$$\mathbf{s}_1 \cdot \mathbf{a}_1 = L_1 x_1 + M_1 y_1 = \left(L_1 + \frac{M_1^2}{L_1}\right)x_1 + \left(y_0 - \frac{M_1}{L_1}x_0\right)$$

$$= \pm a\left[(L_1^2 + M_1^2) - \left\{\frac{M_1 x_0 - L_1 y_0}{a}\right\}^2\right]^{1/2}$$

$$= \pm\frac{a}{n_1}\left[(L_0^2 + M_0^2) - \left\{\frac{M_0 x_0 - L_0 y_0}{a}\right\}^2\right]^{1/2}. \tag{3A.5}$$

This equation immediately leads to Eq. (3.16).

3A.2   Derivation of Eq. (3.24)

We multiply Eq. (3.17) by $n(r)$ to obtain

$$n\frac{d}{ds}\left(n\frac{d}{ds}r\right) - r\left(n\frac{d}{ds}\theta\right)^2 = n\frac{dn}{dr}. \tag{3A.6}$$

On the other hand, from Eq. (3.20),

$$n(r)\frac{d}{ds} = n_0 N_0\frac{d}{dz}. \tag{3A.7}$$

Substituting this into Eq. (3A.6), we obtain

$$n_0 N_0 \frac{d}{dz}\left(n_0 N_0 \frac{dr}{dz}\right) - r\left(n_0 N_0 \frac{d\theta}{dz}\right)^2 = n \frac{dn}{dr}; \qquad (3A.8)$$

hence,

$$\frac{d^2 r}{dz^2} = \frac{1}{2n_0^2 N_0^2} \frac{d(n^2)}{dr} + r\left(\frac{d\theta}{dz}\right)^2. \qquad (3A.9)$$

Therefore, from Eqs. (3A.9) and (3.23),

$$\frac{d^2 r}{dz^2} = \frac{1}{2n_0^2 N_0^2} \frac{d(n^2)}{dr} + \frac{1}{r^3 N_0^2}(x_0 M_0 - y_0 L_0)^2. \qquad (3A.10)$$

We integrate both sides of this equation. From the left-hand side,

$$\int_{r_0}^{r} \frac{d^2 r}{dz^2} dr = \int^{z} \frac{d^2 r}{dz^2}\frac{dr}{dz} dz = \frac{1}{2}\int^{z} \frac{d}{dz}\left(\frac{dr}{dz}\right)^2 dz$$

$$= \frac{1}{2}\left(\frac{dr}{dz}\right)^2 - \frac{1}{2}\left(\frac{dr}{dz}\right)^2_{z=0} = \frac{1}{2}\left(\frac{dr}{dz}\right)^2 - \frac{1}{2}\frac{1 - N_0^2}{N_0^2}, \quad (3A.11)$$

because

$$\left(\frac{dr}{dz}\right)^2_{z=0} = \left(\frac{dx}{dz}\right)^2_{z=0} + \left(\frac{dy}{dz}\right)^2_{z=0} = \frac{M_0^2 + L_0^2}{N_0^2} = \frac{1 - N_0^2}{N_0^2}. \qquad (3A.12)$$

From the right-hand side,

$$\int_{r_0}^{r}[\text{rhs of (3A.10)}]\,dr = \left[\frac{n^2(r)}{2n_0^2 N_0^2} - \frac{1}{2r^2 N_0^2}(x_0 M_0 - y_0 L_0)^2\right]_{r_0}^{r}$$

$$= \frac{1}{2N_0^2}\left[\left\{\frac{n(r)}{n_0}\right\}^2 + \left(1 - \frac{r_0^2}{r^2}\right)\frac{(x_0 M_0 - y_0 L_0)^2}{r_0^2} - 1\right]. \qquad (3A.13)$$

Equating Eqs. (3A.11) and (3A.13), we obtain

$$\left(\frac{dr}{dz}\right)^2 = \frac{1}{N_0^2}\left[\left\{\frac{n(r)}{n_0}\right\}^2 + \left(1 - \frac{r_0^2}{r^2}\right)\frac{(x_0 M_0 - y_0 L_0)^2}{r_0^2} - N_0^2\right]. \quad (3A.14)$$

Equation (3.24) can readily be obtained by integrating the inverse of the square root of Eq. (3A.14),

3A.3 Derivation of Eqs. (3.27) and (3.28)

We substitute Eq. (3.26) into (3.25) and integrate using $\int(a^2 - x^2)^{-1/2}\,dx = \sin^{-1}(x/a)$ to obtain

$$z = (n_0 N_0/\alpha n_1)[\sin^{-1}(r/A) - \sin^{-1}(r_0/A)], \qquad (3A.15)$$

where

$$A = (1/\alpha)[1 - (n_0^2 N_0^2/n_1^2)]^{1/2}. \qquad (3A.16)$$

Rewriting Eq. (3A.15), we obtain Eqs. (3.27) and (3.28).

3A.4 Derivation of Eq. (3.30)

The derivation of Eq. (3.30) will not be difficult if two facts are understood. The first is the vectorial relation between phase constants (wave numbers) in the radial and axial directions and the vectorial relation of the ray direction as shown in Fig. 9.38. (Note that in the present case only meridional rays are considered; hence, the rotational wave number m is zero.)

The second fact is the conservation of the axial wave number β along the path of a specific ray. This conservation can be proved generally for a stratified layer structure as shown in Fig. 3A.1. For example, at the boundary between the second and third layers, from Snell's law,

$$n_2 \sin\theta_2 = n_3 \sin\theta_3. \qquad (3A.17)$$

On the other hand, the phase velocities in these layers are

$$v_2 = c/n_2, \qquad (3A.18a)$$
$$v_3 = c/n_3. \qquad (3A.18b)$$

Hence,

$$\beta_2 = (\omega/v_2)\sin\theta_2 = (\omega/c)n_2 \sin\theta_2, \qquad (3A.19)$$

$$\beta_3 = (\omega/v_3)\sin\theta_3 = (\omega/c)n_3 \sin\theta_3. \qquad (3A.20)$$

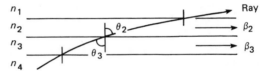

Fig. 3A.1 Proof of the conservation of the axial phase constant.

Equations (3A.17), (3A.19), and (3A.20) lead to $\beta_2 = \beta_3$. This conclusion can be generalized to continuous media.

3A.5   Derivation of Eq. (3.39)

The direct derivation of Eq. (3.39) is possible if we solve Eq. (3.38) assuming that $n(r)$ decreases monotonously with respect to r. In such a derivation, Abel's integral equation is obtained first and is transformed further into Volterra's second-kind integral equation to obtain the solution. However, since this computation is rather tedious, we shall rather "prove" Eq. (3.39).

Substituting Eq. (3.39) into the right-hand side of Eq. (3.38), we immediately obtain

$$C_1[n_1 - n(r_0)] = C_1 n_1 [1 - \text{sech}(\alpha r_0)]. \tag{3A.21}$$

Next, we compute the left-hand side of Eq. (3.38). Substituting Eq. (3.39), we get

$$\int_0^{r_0} [n^2(r) - n^2(r_0)]^{1/2} \, dr$$

$$= \frac{n_1}{\cosh(\alpha r_0)} \int_0^{r_0} \frac{[\cosh^2(\alpha r_0) - \cosh^2(\alpha r)]^{1/2}}{\cosh(\alpha r)} \, dr. \tag{3A.22}$$

By transformation of variables,

$$\sinh(\alpha r) = t, \tag{3A.23a}$$
$$\sinh(\alpha r_0) = a, \tag{3A.23b}$$

Equation (3A.22) becomes

$$\text{Eq. (3A.22)} = \frac{n_1}{\alpha \cosh(\alpha r_0)} \int_0^a \frac{(a^2 - t^2)^{1/2}}{1 + t^2} \, dt. \tag{3A.24}$$

The integral in this equation can be performed by using $y = [(a - t)/(a + t)]^{1/2}$ to give

$$\int_0^a \frac{(a^2 - t^2)^{1/2}}{1 + t^2} \, dt = \frac{\pi}{2} [\cosh(\alpha r_0) - 1]. \tag{3A.25}$$

Substituting Eqs. (3A.21), (3A.22), (3A.24), and (3A.25) into (3.39), we find that (3.39) is the solution of (3.38) provided $\alpha = \pi/2C_1$.

3A.6   Derivation of Eq. (3.49)

From Eqs. (3.47) and (3.48), we obtain

$$1/n_1 = [1 + (r/n)n']^{1/2}/n, \tag{3A.26}$$

where $n' = dn/dr$. By letting $n/n_1 = y$,

$$y = (1 + ry'/y)^{1/2}, \tag{3A.27}$$

$$dr/r = dy/y(y^2 - 1). \tag{3A.28}$$

By replacing y by $t = y^2$, we obtain

$$\frac{dr}{r} = \frac{1}{2}\left(\frac{1}{t-1} - \frac{1}{t}\right)dt, \tag{3A.29}$$

which, under the boundary condition that $y = 1$ at $r = 0$, leads to

$$(\alpha r)^2 = y^{-2} - 1. \tag{3A.30}$$

Equation (3.49) can readily be obtained from this equation.

4A.1   Derivation of Wave Equations in Cylindrical Coordinates

The basic relations between coordinates are

$$x = r\cos\theta, \tag{4A.1}$$

$$y = r\sin\theta, \tag{4A.2}$$

$$r = (x^2 + y^2)^{1/2}, \tag{4A.3}$$

$$\theta = \tan^{-1}(y/x), \tag{4A.4}$$

$$E_r = E_x\cos\theta + E_y\sin\theta \qquad \text{(same for } H_r), \tag{4A.5}$$

$$E_\theta = -E_x\sin\theta + E_y\cos\theta \qquad \text{(same for } H_\theta). \tag{4A.6}$$

From these relations, we obtain

$$\frac{\partial E_z}{\partial x} = \frac{\partial E_z}{\partial r}\frac{\partial r}{\partial x} + \frac{\partial E_z}{\partial \theta}\frac{\partial \theta}{\partial x} = \frac{x}{r}\frac{\partial E_z}{\partial r} - \frac{y}{r^2}\frac{\partial E_z}{\partial \theta}, \tag{4A.7}$$

$$\frac{\partial E_z}{\partial y} = \frac{\partial E_z}{\partial r}\frac{\partial r}{\partial y} + \frac{\partial E_z}{\partial \theta}\frac{\partial \theta}{\partial y} = \frac{y}{r}\frac{\partial E_z}{\partial r} + \frac{x}{r^2}\frac{\partial E_z}{\partial \theta}, \tag{4A.8}$$

$$\frac{\partial^2 E_z}{\partial x^2} = \left(\frac{1}{r} - \frac{x^2}{r^3}\right)\frac{\partial E_z}{\partial r} + \frac{x}{r}\left(\frac{x}{r}\frac{\partial^2 E_z}{\partial r^2} - \frac{y}{r^2}\frac{\partial^2 E_z}{\partial r\,\partial\theta}\right)$$

$$+ \frac{2xy}{r^4}\frac{\partial E_z}{\partial\theta} - \frac{y}{r^2}\left(\frac{x}{r}\frac{\partial^2 E_z}{\partial r\,\partial\theta} - \frac{y}{r^2}\frac{\partial^2 E_z}{\partial\theta^2}\right), \tag{4A.9}$$

$$\frac{\partial^2 E_z}{\partial y^2} = \left(\frac{1}{r} - \frac{y^2}{r^3}\right)\frac{\partial E_z}{\partial r} + \frac{y}{r}\left(\frac{y}{r}\frac{\partial^2 E_z}{\partial r^2} + \frac{x}{r^2}\frac{\partial^2 E_z}{\partial r\,\partial\theta}\right)$$

$$- \frac{2xy}{r^4}\frac{\partial E_z}{\partial\theta} + \frac{x}{r^2}\left(\frac{y}{r}\frac{\partial^2 E_z}{\partial\theta\,\partial r} + \frac{x}{r^2}\frac{\partial^2 E_z}{\partial\theta^2}\right), \tag{4A.10}$$

and also the same relations for magnetic field components. Rewriting the wave equations in Cartesian coordinates by using Eqs. (4A.5)–(4A.10), we obtain the wave equations in cylindrical coordinates.

4A.2   Derivation of Proper Equations for Hybrid Modes under the Weakly Guiding Approximation

We first prepare Bessel-function formulas,

$$J_{n+1}(u) + J_{n-1}(u) = 2(n/u)J_n(u), \tag{4A.11}$$

$$K_{n+1}(w) - K_{n-1}(w) = 2(n/w)K_n(w), \tag{4A.12}$$

$$J_{-n} = (-1)^n J_n, \tag{4A.13}$$

$$K_{-n} = K_n, \tag{4A.14}$$

$$2J'_n = J_{n-1} - J_{n+1}, \tag{4A.15}$$

$$-2K'_n = K_{n-1} + K_{n+1}, \tag{4A.16}$$

from which the following relations are derived [3]:

$$\frac{J'_n(u)}{uJ_n(u)} = \frac{J_{n-1}}{uJ_n} - \frac{n}{u^2} = -\frac{J_{n+1}}{uJ_n} + \frac{n}{u^2}, \tag{4A.17}$$

$$\frac{K'_n(w)}{wK_n(w)} = -\frac{K_{n-1}}{wK_n} - \frac{n}{w^2} = -\frac{K_{n+1}}{wK_n} + \frac{n}{w^2}. \tag{4A.18}$$

Substituting these two equations into Eq. (4.68), we obtain Eqs. (4.70) and (4.71).

5A.1   Derivation of Eqs. (5.7)–(5.12)

Derivation of Eqs. (5.11) and (5.12)

Substituting Eqs. (5.1) and (5.2) into (5.3), we obtain

$$e^{j\omega t} \nabla \times [(\mathbf{E}_t + \mathbf{k}E_z)e^{-j\beta z}] = -j\omega\mu(\mathbf{H}_t + \mathbf{k}H_z)e^{j(\omega t - \beta z)}. \tag{5A.1}$$

By using a vector identity

$$\nabla \times (\mathbf{A}\phi) = \phi \nabla \times \mathbf{A} - \mathbf{A} \times \nabla\phi, \tag{5A.2}$$

we can rewrite the rotation in the left-hand side of Eq. (5A.1) as

$$\nabla \times [(\mathbf{E}_t + \mathbf{k}E_z)e^{-j\beta z}] = e^{-j\beta z}\nabla \times (\mathbf{E}_t + \mathbf{k}E_z) + j\beta e^{-j\beta z}(\mathbf{E}_t + \mathbf{k}E_z) \times \mathbf{k}$$
$$= e^{-j\beta z}(\nabla \times \mathbf{E}_t + E_z\nabla \times \mathbf{k} - \mathbf{k} \times \nabla E_z) + j\beta e^{-j\beta z}[\mathbf{E}_t \times \mathbf{k} + (\mathbf{k}E_z) \times \mathbf{k}]. \tag{5A.3}$$

Since $\nabla \times \mathbf{k} = 0$ and $\mathbf{k} \times \mathbf{k} = 0$, Eqs. (5A.1) and (5A.3) lead to

$$\nabla \times \mathbf{E}_t - \mathbf{k} \times \nabla E_z + j\beta\mathbf{E}_t \times \mathbf{k} = -j\omega\mu(\mathbf{H}_t + \mathbf{k}H_z). \tag{5A.4}$$

On the other hand, since $\mathbf{E}_t$ and E_z are functions of only r and θ,

$$\nabla \times \mathbf{E}_t = \nabla \times (\mathbf{i}E_r + \mathbf{j}E_\theta) = -\frac{\partial E_\theta}{\partial z}\mathbf{i} + \frac{\partial E_r}{\partial z}\mathbf{j} + \frac{1}{r}\left\{\frac{\partial}{\partial r}(rE_\theta) - \frac{\partial E_r}{\partial \theta}\right\}\mathbf{k}$$

$$= \frac{1}{r}\left\{\frac{\partial}{\partial r}(rE_\theta) - \frac{\partial E_r}{\partial \theta}\right\}\mathbf{k}, \tag{5A.5}$$

$$\nabla E_z = \frac{\partial E_z}{\partial r}\mathbf{i} + \frac{1}{r}\frac{\partial E_z}{\partial \theta}\mathbf{j}. \tag{5A.6}$$

These two equations suggest that, as a matter of course, $\nabla \times \mathbf{E}_t$ is parallel to $\mathbf{k}$ whereas ∇E_z is normal to $\mathbf{k}$. Considering these relations, we can divide Eq. (5A.4) into two components normal to each other,

$$\nabla \times \mathbf{E}_t = -j\omega\mu\mathbf{k}H_z, \tag{5A.7}$$

$$-\mathbf{k} \times \nabla E_z + j\beta\mathbf{E}_t \times \mathbf{k} = -j\omega\mu\mathbf{H}_t. \tag{5A.8}$$

Equation (5A.8) can further be rewritten by using a vector identity

$$\mathbf{A} \times (\mathbf{B} \times \mathbf{C}) = (\mathbf{A} \cdot \mathbf{C})\mathbf{B} - (\mathbf{A} \cdot \mathbf{B})\mathbf{C} \tag{5A.9}$$

to give

$$-j\omega\mu\mathbf{k} \times \mathbf{H}_t = -\mathbf{k} \times (\mathbf{k} \times \nabla E_z) + j\beta\mathbf{k} \times (\mathbf{E}_t \times \mathbf{k})$$
$$= -(\mathbf{k} \cdot \nabla E_z)\mathbf{k} + (\mathbf{k} \cdot \mathbf{k})\nabla E_z + j\beta\{(\mathbf{k} \cdot \mathbf{k})\mathbf{E}_t - (\mathbf{k} \cdot \mathbf{E}_t)\mathbf{k}\}$$
$$= \nabla E_z + j\beta\mathbf{E}_t \quad (\text{because} \quad \mathbf{k} \perp \nabla E_z, \quad \mathbf{k} \perp \mathbf{E}_t). \tag{5A.10}$$

Similarly, substituting Eqs. (5.1) and (5.2) into (5.4), we obtain

$$\nabla \times \mathbf{H}_t = j\omega\varepsilon\mathbf{k}E_z, \tag{5A.11}$$

$$j\omega\varepsilon\mathbf{k} \times \mathbf{E}_t = \nabla H_z + j\beta\mathbf{H}_t. \tag{5A.12}$$

Further, substituting Eqs. (5.1) into (5.5) and using

$$\nabla \cdot (\phi\mathbf{A}) = \mathbf{A} \cdot \nabla\phi + \phi\nabla \cdot \mathbf{A}, \tag{5A.13}$$

we obtain

$$0 = \nabla \cdot \left[\varepsilon(\mathbf{E}_t + \mathbf{k}E_z)e^{-j\beta z}\right]$$
$$= (\varepsilon\mathbf{E}_t + \mathbf{k}\varepsilon E_z) \cdot (-j\beta e^{-j\beta z}\mathbf{k}) + e^{-j\beta z}\nabla \cdot (\varepsilon\mathbf{E}_t + \mathbf{k}\varepsilon E_z)$$
$$= -j\beta e^{-j\beta z}\varepsilon E_z + e^{-j\beta z}[\nabla \cdot (\varepsilon\mathbf{E}_t) + \nabla \cdot (\mathbf{k}\varepsilon E_z)]$$
$$= e^{-j\beta z}\{-j\beta\varepsilon E_z + \nabla \cdot (\varepsilon\mathbf{E}_t) + \mathbf{k} \cdot \nabla(\varepsilon E_z) + \varepsilon E_z\nabla \cdot \mathbf{k}\}$$
$$= e^{-j\beta z}\{-j\beta\varepsilon E_z + \nabla \cdot (\varepsilon\mathbf{E}_t)\} \quad [\text{because} \quad \mathbf{k} \perp \nabla(\varepsilon E_z)]. \tag{5A.14}$$

Hence

$$\nabla \cdot (\varepsilon\mathbf{E}_t) = j\beta\varepsilon E_z, \tag{5A.15}$$

and similarly, from Eqs. (5.2) and (5.6),

$$\nabla \cdot (\mu\mathbf{H}_t) = j\beta\mu H_z. \tag{5A.16}$$

We can express $\mathbf{E}_t$ in terms of E_z and H_z if we take the vector product of Eq. (5A.12) and $\mathbf{k}$, and eliminate $\mathbf{k} \times \mathbf{H}_t$ by using (5A.10). First, from Eq. (5A.12),

$$\mathbf{H}_t = (1/j\beta)(j\omega\varepsilon\mathbf{k} \times \mathbf{E}_t - \nabla H_z).$$

Hence, the product with $\mathbf{k}$ is obtained as

$$\mathbf{k} \times \mathbf{H}_t = (1/j\beta)(j\omega\varepsilon\mathbf{k} \times \mathbf{k} \times \mathbf{E}_t - \mathbf{k} \times \nabla H_z)$$
$$= (1/j\beta)(-j\omega\varepsilon\mathbf{E}_t - \mathbf{k} \times \nabla H_z). \tag{5A.17}$$

Substituting Eq. (5A.17) into (5A.10) and performing some computations, we obtain

$$\mathbf{E}_t = \left[-j\beta/(\omega^2\varepsilon\mu - \beta^2)\right][\nabla E_z - (\omega\mu/\beta)\mathbf{k} \times \nabla H_z]. \tag{5A.18}$$

Similarly, from Eqs. (5A.10) and (5A.12),

$$\mathbf{H}_t = [-j\beta/(\omega^2\varepsilon\mu - \beta^2)][(\omega\varepsilon/\beta)\mathbf{k} \times \nabla E_z + \nabla H_z]. \tag{5A.19}$$

Equations (5A.18) and (5A.19) are Eqs. (5.11) and (5.12).

Derivation of Eqs. (5.7) and (5.8)

We first recite Eqs. (5A.11) and (5A.7):

$$E_z = -j(1/\omega\varepsilon)\mathbf{k} \cdot \nabla \times \mathbf{H}_t, \tag{5A.20}$$

$$H_z = j(1/\omega\mu)\mathbf{k} \cdot \nabla \times \mathbf{E}_t. \tag{5A.21}$$

On the other hand, from Eq. (5A.15),

$$\nabla E_z = (1/j\beta)\nabla[\varepsilon^{-1}\nabla \cdot (\varepsilon\mathbf{E}_t)]. \tag{5A.22}$$

Substituting this equation into (5A.10), we obtain

$$-j\omega\mu\mathbf{k} \times \mathbf{H}_t = (1/j\beta)\nabla[\varepsilon^{-1}\nabla \cdot (\varepsilon\mathbf{E}_t)] + j\beta\mathbf{E}_t. \tag{5A.23}$$

Next, we rewrite Eq. (5A.21) as

$$\mu^{-1}\nabla \times \mathbf{E}_t = -j\omega\mathbf{k}H_z, \tag{5A.24}$$

and compute rotations of both sides to obtain

$$\nabla \times (\mu^{-1}\nabla \times \mathbf{E}_t) = -j\omega\nabla \times (\mathbf{k}H_z) = j\omega\mathbf{k} \times \nabla H_z. \tag{5A.25}$$

From Eqs. (5A.23) and (5A.25), we obtain

$$\mu\nabla \times (\mu^{-1}\nabla \times \mathbf{E}_t) - \nabla[\varepsilon^{-1}\nabla \cdot (\varepsilon\mathbf{E}_t)] + \beta^2\mathbf{E}_t$$
$$= j\omega\mu\mathbf{k} \times (\nabla H_z + j\beta\mathbf{H}_t) = j\omega\mu\mathbf{k} \times (j\omega\varepsilon\mathbf{k} \times \mathbf{E}_t) = \omega^2\varepsilon\mu\mathbf{E}_t. \tag{5A.26}$$

Thus, the vectorial wave equation for $\mathbf{E}_r$,

$$-\mu\nabla \times (\mu^{-1}\nabla \times \mathbf{E}_t) + (\omega^2\varepsilon\mu - \beta^2)\mathbf{E}_t + \nabla[\varepsilon^{-1}\nabla \cdot (\varepsilon\mathbf{E}_t)] = 0, \tag{5A.27}$$

is obtained, which is Eq. (5.7).

Similarly, from Eqs. (5A.16), (5A.12), and (5A.20), we obtain the vectorial wave equation for $\mathbf{H}_t$;

$$-\varepsilon\nabla \times (\varepsilon^{-1}\nabla \times \mathbf{H}_t) + (\omega^2\varepsilon\mu - \beta^2)\mathbf{H}_t + \nabla[\mu^{-1}\nabla \cdot (\mu\mathbf{H}_t)] = 0, \tag{5A.28}$$

which is Eq. (5.8).

Derivation of Eqs. (5.9) and (5.10)

Next, we substitute Eq. (5A.18) into (5A.15) to obtain

$$
\varepsilon E_z = -\nabla \cdot \left[\frac{\varepsilon}{\omega^2 \varepsilon\mu - \beta^2} \nabla E_z \right] + \frac{\omega}{\beta} \nabla \cdot \left[\frac{\varepsilon\mu}{\omega^2 \varepsilon\mu - \beta^2} \mathbf{k} \times \nabla H_z \right]
$$

$$
= -\nabla \cdot \left[\frac{\varepsilon}{\omega^2 \varepsilon\mu - \beta^2} \nabla E_z \right] + \frac{\omega}{\beta} \nabla \left(\frac{\varepsilon\mu}{\omega^2 \varepsilon\mu - \beta^2} \right) \cdot (\mathbf{k} \times \nabla H_z)
$$

$$
+ \frac{\omega\varepsilon\mu}{\beta(\omega^2 \varepsilon\mu - \beta^2)} \nabla \cdot (\mathbf{k} \times \nabla H_z), \tag{5A.29}
$$

where

$$
\nabla \cdot (\mathbf{k} \times \nabla H_z) = -\mathbf{k} \cdot \nabla \times (\nabla H_z) = 0.
$$

Therefore, from Eq. (5A.29),

$$
\frac{\omega^2 \varepsilon\mu - \beta^2}{\varepsilon} \nabla \cdot \left[\frac{\varepsilon}{\omega^2 \varepsilon\mu - \beta^2} \nabla E_z \right] + (\omega^2 \varepsilon\mu - \beta^2) E_z
$$

$$
- \frac{\omega^2 \varepsilon\mu - \beta^2}{\varepsilon\mu} \nabla \left(\frac{\varepsilon\mu}{\omega^2 \varepsilon\mu - \beta^2} \right) \cdot \left(\mathbf{k} \times \frac{\omega\mu}{\beta} \nabla H_z \right) = 0. \tag{5A.30}
$$

Furthermore, using a vector identity

$$
\nabla[\ln \phi(r, \theta)] = \mathbf{i} \frac{\partial}{\partial r} \ln \phi + \mathbf{j} \frac{1}{r} \frac{\partial}{\partial \theta} \ln \phi
$$

$$
= \mathbf{i} \frac{1}{\phi} \frac{\partial \phi}{\partial r} + \mathbf{j} \frac{1}{\phi} \frac{1}{r} \frac{\partial \phi}{\partial \theta} = \frac{1}{\phi} \nabla \phi, \tag{5A.31}
$$

we can write

$$
\frac{\omega^2 \varepsilon\mu - \beta^2}{\varepsilon\mu} \nabla \left(\frac{\varepsilon\mu}{\omega^2 \varepsilon\mu - \beta^2} \right) = \nabla \left[\ln \frac{\varepsilon\mu}{\omega^2 \varepsilon\mu - \beta^2} \right], \tag{5A.32}
$$

with which Eq. (5A.30) can be rewritten as

$$
\frac{\omega^2 \varepsilon\mu - \beta^2}{\varepsilon} \nabla \cdot \left[\frac{\varepsilon}{\omega^2 \varepsilon\mu - \beta^2} \nabla E_z \right] + (\omega^2 \varepsilon\mu - \beta^2) E_z
$$

$$
- \nabla \left[\ln \frac{\varepsilon\mu}{\omega^2 \varepsilon\mu - \beta^2} \right] \cdot \left(\mathbf{k} \times \frac{\omega\mu}{\beta} \nabla H_z \right) = 0. \tag{5A.33}
$$

This is Eq. (5.9).

Similarly, substituting Eq. (5A.19) into (5A.16), we obtain

$$\frac{\omega^2 \varepsilon \mu - \beta^2}{\mu} \nabla \cdot \left[\frac{\mu}{\omega^2 \varepsilon \mu - \beta^2} \nabla H_z \right] + (\omega^2 \varepsilon \mu - \beta^2) H_z$$

$$+ \nabla \left[\ln \frac{\varepsilon \mu}{\omega^2 \varepsilon \mu - \beta^2} \right] \cdot \left(\mathbf{k} \times \frac{\omega \varepsilon}{\beta} \nabla E_z \right) = 0, \tag{5A.34}$$

which is Eq. (5.10).

5A.2   Derivation of Eq. (5.23)

If we assume that the θ-dependence of E_r and E_θ are both expressed as $\exp(-jn\theta)$, Eqs. (5.19) and (5.20) are rewritten as

$$\frac{d^2 E_r}{dr^2} + \frac{1}{r} \frac{dE_r}{dr} + \left[\omega^2 \varepsilon(r)\mu_0 - \beta^2 - \frac{n^2 + 1}{r^2} \right] E_r + j \frac{2n}{r^2} E_\theta = 0, \tag{5A.35}$$

$$\frac{d^2 E_\theta}{dr^2} + \frac{1}{r} \frac{dE_\theta}{dr} + \left[\omega^2 \varepsilon(r)\mu_0 - \beta^2 - \frac{n^2 + 1}{r^2} \right] E_\theta - j \frac{2n}{r} E_r = 0. \tag{5A.36}$$

In these equations, as well as in Eqs. (5.19) and (5.20), E_r and E_θ are not separable; i.e., these equations are not scalar wave equations.

Kurtz and Streifer (see [1] in Chapter 5) found that if

$$E_r^{(i)} = \pm jE_\theta^{(i)} \qquad (i = 1, 2), \tag{5A.37}$$

we can obtain separate equations for E_r and E_θ. The upper and lower signs correspond to $i = 1$ and $i = 2$, respectively. [The case $i = 1$ corresponds to HE modes, whereas $i = 2$ corresponds to TE, TM, and EM modes. See Eq. (5.79).] Substituting Eq. (5A.37) into Eqs. (5A.35) and (5A.36), we obtain

$$\frac{d^2 E_b^{(i)}}{dr^2} + \frac{1}{r} \frac{dE_b^{(i)}}{dr} + \left[\omega^2 \varepsilon(r)\mu_0 - \beta^2 - \frac{(n \mp 1)^2}{r^2} \right] E_b^{(i)} = 0, \tag{5A.38}$$

$$E_b^{(i)} = E_r^{(i)} \quad \text{or} \quad E_\theta^{(i)} \qquad (i = 1, 2).$$

Therefore, by using new functions $R^{(i)}$, $E_r^{(i)}$ and $E_\theta^{(i)}$ can be expressed as

$$E_r^{(i)} = \pm jR^{(i)}(r)e^{-jn\theta} \tag{5A.39}$$

$$E_\theta^{(i)} = R^{(i)}(r)e^{-jn\theta} \qquad (i = 1, 2), \tag{5A.40}$$

where $R^{(i)}$ is the solution of scalar wave equations;

$$\frac{d^2 R^{(i)}}{dr^2} + \frac{1}{r}\frac{dR^{(i)}}{dr} + \left[\omega^2\varepsilon(r)\mu_0 - \beta^2 - \frac{(n \mp 1)^2}{r^2}\right]R^{(i)} = 0 \qquad (i = 1, 2). \quad (5A.41)$$

5A.3   Definitions and Properties of Airy Functions Ai(x) and Bi(x)

Definitions and some important properties will be cited from Ref. [7] of Chapter 5.

(1)   The solutions of a differential equation

$$d^2 y/dx^2 + xy = 0 \qquad (5A.42)$$

are called Airy functions. Two linearly independent solutions are found, and are denoted Ai(x) and Bi(i).

(2)   The functions Ai(x) and Bi(x) are

$$Ai(x) = C_1 f(x) - C_2 g(x), \qquad (5A.43)$$
$$Bi(x) = \sqrt{3}[C_1 f(x) + C_2 g(x)], \qquad (5A.44)$$

where

$$f(x) = 1 + \frac{1}{3!}x^3 + \frac{1\cdot 4}{6!}x^6 + \frac{1\cdot 4\cdot 7}{9!}x^9 + \cdots = \sum_0^\infty 3^k\left(\frac{1}{3}\right)_k \frac{x^{3k}}{(3k)!},$$

$$(5A.45)$$

$$g(x) = x + \frac{2}{4!}x^4 + \frac{2\cdot 5}{7!}x^7 + \frac{2\cdot 5\cdot 8}{10!}x^{10} + \cdots = \sum_0^\infty 3^k\left(\frac{2}{3}\right)_k \frac{x^{3k+1}}{(3k+1)!},$$

$$(5A.46)$$

$$C_1 = 3^{-2/3}/\Gamma(\tfrac{2}{3}) = 0.35502, \qquad (5A.47)$$
$$C_2 = 3^{-1/3}/\Gamma(\tfrac{1}{3}) = 0.25881. \qquad (5A.48)$$

(3)   The functions Ai(x), Bi(x) and their derivatives Ai$'(x)$, Bi$'(x)$ are shown in Fig. 5A.1. The values at $x = 0$ are

$$Ai(0) = Bi(0)/\sqrt{3} = C_1, \qquad (5A.49)$$
$$-Ai'(0) = Bi'(0)/\sqrt{3} = C_2. \qquad (5A.50)$$

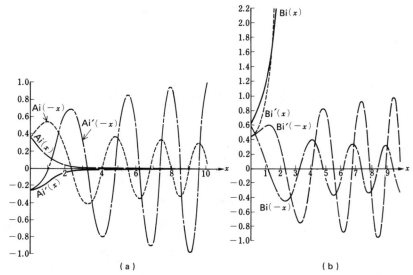

Fig. 5A.1   Airy functions and their derivatives: (a) Ai($\pm x$) and Ai$'(\pm x)$, (b) Bi($\pm x$) and Bi$'(\pm x)$.

5A.4   Derivation of Eqs. (5.136)–(5.139) (Determination of Constants A–G, φ, and β_1)

(1)   For $r < r_1$, Eq. (5.119) is cited:

$$\hat{R} = Aq^{-1/4} \exp\left[-\beta_1 \int_r^{r_1} \sqrt{q(r)}\, dr \right] \qquad (5A.51)$$

(2)   For $r \lesssim r_1$, from the diverging property of Bi(x) shown in Fig. 5A.1, we must assume in Eq. (5.124) that

$$C = 0. \qquad (5A.52)$$

Therefore, substituting Eq. (5.132) into (5.124), we obtain

$$\hat{R} = Bq^{-1/4}\xi'^{1/4}(1/2\sqrt{\pi})\xi'^{-1/4} \exp(-\tfrac{2}{3}\xi'^{3/2}). \qquad (5A.53)$$

Equating Eqs. (5A.51) and (5A.53) and using (5.125), we obtain

$$A = B/2\sqrt{\pi}. \qquad (5A.54)$$

(3)   For $r_1 \lesssim r$, we obtain, substituting Eqs. (5A.52) and (5.134) into (5.126),

$$\hat{R} = Bp^{-1/4}\xi^{1/4}(1/\sqrt{\pi})\xi^{-1/4} \sin(\tfrac{2}{3}\xi^{3/2} + \tfrac{1}{4}\pi). \qquad (5A.55)$$

(4) For $r_1 < r < r_2$, Eq. (5.117) is cited:

$$\hat{R} = Dp^{-1/4} \sin\left[\beta_1 \int_{r_1}^{r} \sqrt{\rho(r)}\, dr + \varphi\right].$$ (5A.56)

Equating Eqs. (5A.55) and (5A.56) and using (5.127), we obtain

$$D = B/\sqrt{\pi},$$ (5A.57)

$$\varphi = \pi/4.$$ (5A.58)

(5) For $r \lesssim r_2$, since $F = 0$ [see (6)], substituting Eq. (5.134) and $F = 0$ into (5.128), we obtain

$$\hat{R} = Ep^{-1/4}\eta^{1/4}(1/\sqrt{\pi})\eta^{-1/4}\sin(\tfrac{2}{3}\eta^{3/2} + \tfrac{1}{4}\pi).$$ (5A.59)

Equating Eqs. (5A.56) and (5A.59) and using (5.129), we obtain

$$D\sin\left[\beta_1 \int_{r_1}^{r}\sqrt{p(r)}\, dr + \tfrac{1}{4}\pi\right] = (E/\sqrt{\pi})\sin\left[\beta_1 \int_{r}^{r_2}\sqrt{p(r)}\, dr + \tfrac{1}{4}\pi\right].$$ (5A.60)

(6) For $r_2 \lesssim r$, from the diverging properties of $\mathrm{Bi}(x)$ shown in Fig. 5A.1, we must assume in Eq. (5.130) that

$$F = 0.$$ (5A.61)

Therefore, substituting Eq. (5.132) into (5.130), we obtain

$$\hat{R} = Eq^{-1/4}\eta'^{1/4}(1/2\sqrt{\pi})\eta'^{-1/4}\exp(-\tfrac{2}{3}\eta'^{3/2}).$$ (5A.62)

(7) For $r_2 < r$, Eq. (5.120) is cited:

$$\hat{R} = Gq^{-1/4}\exp\left[-\beta_1 \int_{r_2}^{r}\sqrt{q(r)}\, dr\right].$$ (5A.63)

Equating Eqs. (5A.62) and (5A.63) and using (5.131), we obtain

$$G = E/2\sqrt{\pi}.$$ (5A.64)

Thus, all the relations in Eqs. (5.136) and (5.137) have been derived.

(8) Equations (5.138) and (5.139) are derived from (5.137). Generally, for $P\sin X(r) = Q\sin Y(r)$ to hold regardless of r, either condition

(i) $P = Q$ and $X(r) + Y(r) = l\pi$, where l is an odd integer, or
(ii) $P = -Q$ and $X(r) + Y(r) = l'\pi$, where l' is an even integer,

is required. Using these relations, Eqs. (5.138) and (5.139) can be derived.

5A.5   Derivation of Eq. (5.193)

Substituting Eq. (5.190) into the first term of (5.181), we obtain

$$\int_0^a \left[\left(\frac{dR}{dr} \right)^2 + \frac{m^2}{r^2} R^2 \right] r\, dr = \sum_{k=1}^\infty \sum_{l=1}^\infty C_k C_l \int_0^a \left[\frac{dF_{mk}}{dr} \frac{dF_{ml}}{dr} + \frac{m^2}{r^2} F_{mk} F_{ml} \right] r\, dr.$$

(5A.65)

By partially integrating the right-hand side of this equation, we rewrite it as

$$\int_0^a \left[\frac{dF_{mk}}{dr} \frac{dF_{ml}}{dr} + \frac{m^2}{r^2} F_{mk} F_{ml} \right] r\, dr = \left[F_{ml} r \frac{dF_{mk}}{dr} \right]_0^a$$

$$- \int_0^a F_{ml} \left\{ \frac{1}{r} \frac{d}{dr} \left(r \frac{dF_{mk}}{dr} \right) - \frac{m^2}{r^2} F_{mk} \right\} r\, dr. \qquad (5A.66)$$

Since λ_k is the root of Eq. (5.189),

$$\left[\frac{dF_{mk}}{dr} \right]_{r=a} = \frac{\lambda_k J'_m(\lambda_k)}{a J_m(\lambda_k)} = \frac{\Phi_\beta}{a}. \qquad (5A.67)$$

On the other hand, from Eq. (5.188) and the Bessel-function formulas,

$$\frac{1}{r} \frac{d}{dr} \left(r \frac{dF_{mk}}{dr} \right) - \frac{m^2}{r^2} F_{mk} = \frac{1}{J_m(\lambda_k)} \left\{ \frac{1}{r} \frac{d}{dr} \left[r \frac{dJ_m(\lambda_k r/a)}{dr} \right] - \frac{m^2}{r^2} J_m(\lambda_k r/a) \right\}$$

$$= -\left(\frac{\lambda_k}{a} \right)^2 \frac{J_m(\lambda_k r/a)}{J_m(\lambda_k)} = -\left(\frac{\lambda_k}{a} \right)^2 F_{mk}. \qquad (5A.68)$$

Hence, Eq. (5A.66) is rewritten, by using (5A.67) and (5A.68), as

$$\text{Eq. (5A.66)} = \Phi_\beta + \left(\frac{\lambda_k}{a} \right)^2 \int_0^a F_{mk} F_{ml} r\, dr$$

$$= \Phi_\beta + \frac{\lambda_k^2}{2} [1 + (\Phi_\beta^2 - m^2)/\lambda_k^2] \delta_{kl}. \qquad (5A.69)$$

Therefore, Eq. (5A.65) now becomes

$$\int_0^a \left[\left(\frac{dR}{dr} \right)^2 + \frac{m^2}{r^2} R^2 \right] r\, dr = \sum_{k=1}^\infty \sum_{l=1}^\infty C_k C_l \{ \Phi_\beta + \tfrac{1}{2} \lambda_k^2 [1 + (\Phi_\beta^2 - m^2)/\lambda_k^2] \delta_{kl} \}.$$

(5A.70)

Next, the second term in Eq. (5.181) becomes

$$\int_0^a [\omega^2\varepsilon(r)\mu_0 - \beta^2]R^2 r\,dr = (\omega^2\varepsilon_1\mu_0 - \beta^2)\int_0^a R^2 r\,dr - \omega^2\varepsilon_1\mu_0\int_0^a f(r)R^2 r\,dr$$

$$= (\omega^2\varepsilon_1\mu_0 - \beta^2)\sum_{k=1}^{\infty}\sum_{l=1}^{\infty} C_k C_l \int_0^a F_{mk}F_{ml}r\,dr$$

$$- \omega^2\varepsilon_1\mu_0\sum_{k=1}^{\infty}\sum_{l=1}^{\infty} C_k C_l \int_0^a f(r)F_{mk}F_{ml}r\,dr.$$

$$(5A.71)$$

If we here use $(\omega^2\varepsilon_1\mu_0 - \beta^2) = u^2/a^2$ and the orthogonality relation [Eq. (5.191)] and define

$$A_{mkl} = (1/\Delta a^2)\int_0^a f(r)F_{mk}F_{ml}r\,dr, \qquad (5A.72)$$

Eq. (5A.71) can further be rewritten as

$$\int_0^a [\omega^2\varepsilon(r)\mu_0 - \beta^2]R^2 r\,dr = \tfrac{1}{2}\mu^2\sum_{k=1}^{\infty}\sum_{l=1}^{\infty} C_k C_l[1 + (\Phi_\beta^2 - m^2)/\lambda_k^2]\delta_{kl}$$

$$- \tfrac{1}{2}v^2\sum_{k=1}^{\infty}\sum_{l=1}^{\infty} C_k C_l A_{mkl}. \qquad (5A.73)$$

Substituting Eqs. (5A.70) and (5A.73) into (5.181), we obtain the functional

$$I[C_1, C_2, \ldots] = \tfrac{1}{2}\sum_{k=1}^{\infty}\sum_{l=1}^{\infty} C_k C_l\{(\lambda_k^2 - u^2)[1 + (\Phi_\beta^2 - m^2)/\lambda_k^2]\delta_{kl} + v^2 A_{mkl}\}.$$

$$(5A.74)$$

5A.6   Proof of Eq. (5.198)

When $R(r)$ is the solution of the scalar wave equation (5.178) satisfying the boundary conditions (5.179) and (5.180),

$$\beta^2 = \frac{\int_0^\infty k^2 n^2(r)R^2 r\,dr - \int_0^\infty \{(dR/dr)^2 + (m^2/r^2)R^2\}r\,dr}{\int_0^\infty R^2 r\,dr}, \qquad (5A.75)$$

and is stationary with respect to small variations of $R(r)$. The proof follows.

When $R(r)$ varies slightly by $\delta R(r)$, the resultant variation of Eq. (5A.75) is expressed as

$$\tfrac{1}{2}\delta\beta^2 \int_0^\infty R^2 r\,dr = -\beta^2 \int_0^\infty R\,\delta R r\,dr + \int_0^\infty k^2 n^2 R\,\delta R r\,dr$$

$$- \int_0^\infty \left[\frac{dR}{dr}\frac{d\,\delta R}{dr} + \frac{m^2}{r^2} R\,\delta R\right] r\,dr. \qquad (5A.76)$$

On integrating by parts the third term of this equation, we obtain

$$\int_0^\infty \left[\frac{dR}{dr}\frac{d\,\delta R}{dr} + \frac{m^2}{r^2} R\,\delta R\right] r\,dr$$

$$= \left[r\frac{dR}{dr}\delta R\right]_0^\infty - \int_0^\infty \delta R\left\{\frac{1}{r}\frac{d}{dr}\left(r\frac{dR}{dr}\right) - \frac{m^2}{r^2} R\right\} r\,dr. \qquad (5A.77)$$

Substituting Eq. (5A.77) into (5A.76), we obtain

$$\tfrac{1}{2}\delta\beta^2 \int_0^\infty R^2 r\,dr = -\left[r\frac{dR}{dr}\delta R\right]_0^\infty + \int_0^\infty \delta R\left\{\frac{1}{r}\frac{d}{dr}\left(r\frac{dR}{dr}\right)\right.$$

$$\left. + \left[k^2 n^2 - \beta^2 - \frac{m^2}{r^2}\right] R\right\} r\,dr. \qquad (5A.78)$$

Since δR is arbitrary, when $R(r)$ satisfies Eqs. (5.178)–(5.180) and

$$\lim_{r\to\infty} r\,dR/dr = 0, \qquad (5A.79)$$

the right-hand side of Eq. (5A.78) becomes zero. Thus, β^2 given as Eq. (5A.75) is stationary, and this β is equal to that used in Eq. (5.178).

5A.7   Derivation of Eq. (5.199)

We assume that when the wave number k varies slightly, β and R become $\beta + \delta\beta$ and $R + \delta R$, respectively. In this case, from Eq. (5.198),

$$(\beta + \delta\beta)^2 \int_0^\infty (R + \delta R)^2 r\,dr - \int_0^\infty (kn + \delta(kn)]^2 (R + \delta R)^2 r\,dr$$

$$+ \int_0^\infty \left\{\left[\frac{d(R + \delta R)}{dr}\right]^2 + \frac{m^2}{r^2}(R + \delta R)^2\right\} r\,dr = 0. \qquad (5A.80)$$

Neglecting the δ^2 terms, we can rewrite this equation as

$$\left\{ \beta^2 \int_0^\infty R^2(r) r\, dr - \int_0^\infty [kn(r)]^2 R^2(r) r\, dr + \int_0^\infty \left[\left(\frac{dR}{dr} \right)^2 + \frac{m^2}{r^2} R^2 \right] r\, dr \right\}$$

$$+ \left\{ \beta^2 \int_0^\infty \delta R^2 r\, dr - \int_0^\infty k^2 n^2\, \delta R^2 r\, dr + \int_0^\infty \delta \left[\left(\frac{dR}{dr} \right)^2 + \frac{m^2}{r^2} R^2 \right] r\, dr \right\}$$

$$+ 2 \left\{ \beta\, \delta\beta \int_0^\infty R^2(r) r\, dr - \int_0^\infty kn(r)\, \delta[kn(r)] R^2(r) r\, dr \right\} = 0. \qquad (5A.81)$$

Substituting Eq. (5.198) into the first term of this equation, and also substituting the following relation obtained from the stationary condition $\delta\beta^2 = 0$,

$$\beta^2 \int_0^\infty \delta R^2 r\, dr - \int_0^\infty k^2 n^2\, \delta R^2 r\, dr + \int_0^\infty \delta \left[\left(\frac{dR}{dr} \right)^2 + \frac{m^2}{r^2} R^2 \right] r\, dr = 0,$$

$$(5A.82)$$

into the second term of Eq. (5A.81), we obtain

$$\beta\, \delta\beta \int_0^\infty R^2(r) r\, dr - \int_0^\infty kn\, \delta(kn) R^2(r) r\, dr = 0. \qquad (5A.83)$$

This can be rewritten as

$$\beta \frac{d\beta}{dk} = \frac{\int_0^\infty kn[d(kn)/dk]\, R^2(r) r\, dr}{\int_0^\infty R^2(r) r\, dr}. \qquad (5A.84)$$

5A.8 Derivation of Eq. (5.200)

Equation (5.199) can also be expressed as

$$\beta \frac{d\beta}{dk} = \frac{\int_0^\infty \frac{1}{2}[d(k^2 n^2)/dk]\, R^2 r\, dr}{\int_0^\infty R^2 r\, dr}. \qquad (5A.85)$$

As given in Eq. (5.101), the refractive-index profile in the core is

$$n^2 = n_1^2[1 - f(r)]. \qquad (5A.86)$$

We here express the function $f(r)$ as a product of a function of the wavelength Δ (which express the effect of the material dispersion) and a function of only the radial coordinate $g(r)$, so that

$$f(r) = 2 \Delta g(r). \tag{5A.87}$$

From Eqs. (5A.86) and (5A.87), since $g(r)$ is independent of k,

$$\frac{1}{2} \frac{d(k^2 n^2)}{dk} = \frac{1}{2} \frac{d}{dk} \left[k^2 n_1^2 (1 - 2 \Delta g) \right]$$

$$= k n_1 \frac{d(k n_1)}{dk} (1 - 2 \Delta g) - k^2 n_1^2 \frac{d\Delta}{dk} g. \tag{5A.88}$$

If we define and use

$$N_1 = d(k n_1)/dk, \tag{5A.89}$$

$$y = (2 n_1 / N_1)(k/\Delta) \, d\Delta/dk, \tag{5A.90}$$

we can further rewrite Eq. (5A.88) as

$$\tfrac{1}{2} d(k^2 n^2)/dk = k n_1 N_1 \left[1 - 2 \Delta (1 + \tfrac{1}{4} y) g(r) \right] = k n_1 N_1 \left[1 - (1 + \tfrac{1}{4} y) f(r) \right]. \tag{5A.91}$$

Substituting Eq. (5A.91) into (5A.85), we obtain

$$\frac{\beta}{k n_1} \frac{d\beta}{dk} = N_1 \left\{ 1 - (1 + \tfrac{1}{4} y) \frac{\int_0^\infty f(r) R^2 r \, dr}{\int_0^\infty R^2 r \, dr} \right\}. \tag{5A.92}$$

Since $R(r)$ can be expressed as

$$R(r) = \begin{cases} \displaystyle\sum_{k=1}^\infty C_k F_{mk}(r) & (0 \leqq r \leqq a), \\[4mm] R(a) \dfrac{K_m(wr/a)}{K_m(w)} & (a < r), \end{cases} \tag{5A.93}$$

the integral in the denominator of Eq. (5A.92) is

$$\int_0^\infty R^2 r \, dr = \sum_{k=1}^\infty \sum_{l=1}^\infty C_k C_l \int_0^a F_{mk} F_{ml} r \, dr + R^2(a) \int_a^\infty \left[\frac{K_m(wr/a)}{K_m(w)} \right]^2 r \, dr$$

$$= \sum_{k=1}^\infty \sum_{l=1}^\infty C_k C_l \frac{a^2}{2} \left[1 + (\Phi_\beta^2 - m^2)/\lambda_k^2 \right] \delta_{kl}$$

$$+ R^2(a) \frac{a^2}{2} \left[\frac{K_{m-1}(w) K_{m+1}(w)}{K_m^2(w)} - 1 \right]. \tag{5A.94}$$

On the other hand, from Eq. (5A.93),

$$R(a) = \sum_{k=1}^{\infty} C_k. \tag{5A.95}$$

Therefore, Eq. (5A.94) becomes

$$\int_0^{\infty} R^2 r\, dr = \frac{a^2}{2} \sum_{k=1}^{\infty} \sum_{l=1}^{\infty} C_k C_l \left\{ [1 + (\Phi_\beta^2 - m^2)/\lambda_k^2] \delta_{kl} + \left(\frac{1}{\xi_m(w)} - 1 \right) \right\}, \tag{5A.96}$$

where

$$\xi_m(w) = K_m^2(w)/K_{m-1}(w) K_{m+1}(w). \tag{5A.97}$$

Next, we compute the other integral in Eq. (5A.92). Since $f(r) = 2\Delta$ (const) in the cladding,

$$\int_0^{\infty} f(r) R^2 r\, dr = \sum_{k=1}^{\infty} \sum_{l=1}^{\infty} C_k C_l \int_0^a f(r) F_{mk} F_{ml} r\, dr$$

$$+ R^2(a) 2\Delta \int_a^{\infty} \left[\frac{K_m(wr/a)}{K_m(w)} \right]^2 r\, dr$$

$$= \Delta a^2 \sum_{k=1}^{\infty} \sum_{l=1}^{\infty} C_k C_l A_{mkl} + R^2(a) \Delta a^2 \left[\frac{1}{\xi_m(w)} - 1 \right]$$

$$= \Delta a^2 \sum_{k=1}^{\infty} \sum_{l=1}^{\infty} C_k C_l \left[A_{mkl} + \frac{1}{\xi_m(w)} - 1 \right]. \tag{5A.98}$$

Substituting Eqs. (5A.96) and (5A.98) into (5A.92), we obtain

$$\frac{\beta}{kn_1} \frac{d\beta}{dk} = N_1 \left\{ 1 - (2 + \tfrac{1}{2}y) \right.$$

$$\left. \times \Delta \frac{\sum_{k=1}^{\infty} \sum_{l=1}^{\infty} C_k C_l [A_{mkl} + 1/\xi_m - 1]}{\sum_{k=1}^{\infty} \sum_{l=1}^{\infty} C_k C_l [\{1 + (\Phi_\beta^2 - m^2)/\lambda_k^2\} \delta_{kl} + 1/\xi_m - 1]} \right\}. \tag{5A.99}$$

5A.9 Table of Exact Normalized Cutoff Frequencies

Table 5A.1 shows normalized cutoff frequencies computed by the series-expansion method for $\alpha = 1, 2, 4$, and 10, $\rho = 1$, $l = 1$–10, $m = 0$–9 (after Oyamada and Okoshi (see [13] in Chapter 5). For most cases the relative computation error is estimated to be less than 5×10^{-5} (see p. 126).

Part I (m = 0–4)

(A) ALPHA = 1

	M = 0	M = 1	M = 2	M = 3	M = 4
L = 1	0.000000	4.381552	7.218053	9.918859	12.569812
L = 2	5.948312	8.933073	11.714840	14.414960	17.073726
L = 3	10.773218	13.575572	16.301533	18.981256	21.632252
L = 4	15.535689	18.247972	20.930571	23.587745	26.226232
L = 5	20.277313	22.934087	25.583111	28.219431	30.844425
L = 6	25.009182	27.627715	30.250148	32.868007	35.479889
L = 7	29.735625	32.325942	34.926809	37.528495	40.128122
L = 8	34.458707	37.027206	39.610216	42.197722	44.786075
L = 9	39.179548	41.730590	44.298550	46.873568	49.451613
L =10	43.898814	46.435519	48.990603	51.554558	54.123189

(B) ALPHA = 2

	M = 0	M = 1	M = 2	M = 3	M = 4
L = 1	0.000000	3.518050	5.743923	7.847594	9.904203
L = 2	5.067506	7.451446	9.645060	11.759842	13.833188
L = 3	9.157606	11.423744	13.590328	15.702133	17.780248
L = 4	13.197225	15.408067	17.554802	19.660836	21.739339
L = 5	17.220229	19.397792	21.529546	23.629565	25.706695
L = 6	21.235517	23.390452	25.510500	27.604910	29.679959
L = 7	25.246531	27.384901	29.495527	31.584878	33.657593
L = 8	29.254906	31.380530	33.483388	35.568216	37.638559
L = 9	33.261524	35.376982	37.473309	39.554097	41.622126
L =10	37.266908	39.374036	41.464782	43.541950	45.607769

(C) ALPHA = 4

	M = 0	M = 1	M = 2	M = 3	M = 4
L = 1	0.000000	2.999553	4.849936	6.580971	8.263079
L = 2	4.555590	6.609370	8.464058	10.231664	11.951132
L = 3	8.226613	10.202567	12.058915	13.846820	15.590939
L = 4	11.855771	13.795271	15.650208	17.449592	19.210069
L = 5	15.470632	17.388235	19.240901	21.047174	22.818915
L = 6	19.078758	20.981466	22.831718	24.642327	26.422112
L = 7	22.683119	24.574905	26.422839	28.236268	30.021978
L = 8	26.285139	28.168503	30.014287	31.829588	33.619775
L = 9	29.885592	31.762224	33.606037	35.422593	37.216237
L =10	33.484940	35.356041	37.198055	39.015451	40.811810

(D) ALPHA =10

	M = 0	M = 1	M = 2	M = 3	M = 4
L = 1	0.000000	2.649259	4.242970	5.713952	7.130066
L = 2	4.174340	6.026765	7.675864	9.228729	10.724173
L = 3	7.591270	9.378183	11.042524	12.632376	14.171512
L = 4	10.956339	12.718355	14.390161	16.001372	17.568038
L = 5	14.309667	16.054404	17.730210	19.355070	20.940528
L = 6	17.653808	19.388406	21.066491	22.700703	24.299729
L = 7	20.993947	22.721218	24.400646	26.041594	27.650827
L = 8	24.331596	26.053267	27.733492	29.379478	30.996638
L = 9	27.667573	29.384796	31.065481	32.715347	34.338818
L =10	31.002366	32.715950	34.396882	36.049806	37.678405

Part II (m = 5–9)

(A) ALPHA = 1

	M = 5	M = 6	M = 7	M = 8	M = 9
L = 1	15.197459	17.812875	20.421423	23.025945	25.628033
L = 2	19.709153	22.330576	24.943276	27.550414	30.153954
L = 3	24.264756	26.885041	29.497118	32.103625	34.706342
L = 4	28.851316	31.466823	34.075488	36.679279	39.279614
L = 5	33.460362	36.069329	38.673037	41.272834	43.869767
L = 6	38.086220	40.687894	43.285842	45.880906	48.473796
L = 7	42.724979	45.319145	47.910979	50.500916	53.089385
L = 8	47.373859	49.960595	52.546222	55.130863	57.714716
L = 9	52.030832	54.610369	57.189855	59.769165	62.348452
L =10	56.694386	59.267010	61.840825	64.411383	67.028140

[a] After K. Oyamoto and T. Okoshi, *IEEE Trans. Microwave Theory Tech. MTT-28*, No. 10, 1113–1118 (1980)

Table 5A.1

Part II (m = 5–9) (cont.)

(B)   ALPHA = 2

	M = 5	M = 6	M = 7	M = 8	M = 9
L = 1	11.937778	13.958687	15.972134	17.980980	19.986899
L = 2	15.882132	17.915717	19.939211	21.955873	23.967812
L = 3	19.836252	21.877116	23.907310	25.929824	27.946725
L = 4	23.798431	25.843439	27.878016	29.904747	31.925509
L = 5	27.766848	29.814148	31.851566	33.881301	35.905008
L = 6	31.740089	33.788551	35.827781	37.859640	39.885574
L = 7	35.717108	37.766029	39.806371	41.839717	43.867326
L = 8	39.697134	41.746066	43.787038	45.821404	47.850265
L = 9	43.679590	45.728244	47.769510	49.804555	51.834340
L =10	47.664039	49.712229	51.753553	53.789020	55.819476

(C)   ALPHA = 4

	M = 5	M = 6	M = 7	M = 8	M = 9
L = 1	9.919787	11.561690	13.194344	14.820944	16.443449
L = 2	13.640769	15.310610	16.966698	18.612914	20.251867
L = 3	17.305095	18.997798	20.674632	22.339436	23.994945
L = 4	20.942088	22.652626	24.346560	26.027414	27.697811
L = 5	24.564169	26.288640	27.996502	29.690900	31.374249
L = 6	28.177403	29.912895	31.632152	33.337945	35.032461
L = 7	31.785053	33.529395	35.258060	36.973481	38.677623
L = 8	35.389006	37.140562	38.877075	40.600686	42.313156
L = 9	38.990421	40.747936	42.491064	44.221694	45.941406
L =10	42.590040	44.352537	46.101294	47.837985	49.564030

(D)   ALPHA =10

	M = 5	M = 6	M = 7	M = 8	M = 9
L = 1	8.515250	9.880903	11.233353	12.576461	13.912734
L = 2	12.181310	13.611131	15.020595	16.414397	17.795860
L = 3	15.673698	17.147726	18.599600	20.033620	21.452985
L = 4	19.100437	20.605606	22.088605	23.553210	25.002323
L = 5	22.494501	24.022693	25.529368	27.017809	28.490604
L = 6	25.869832	27.415705	28.940963	30.448461	31.940500
L = 7	29.233415	30.793270	32.333486	33.856557	35.364526
L = 8	32.589152	34.160326	35.712832	37.248857	38.770223
L = 9	35.939396	37.519910	39.082683	40.629648	42.162433
L =10	39.285653	40.873995	42.445467	44.001788	45.544419

5A.10   Proof of the Validity of the Functional [Eq. (5.242)]

We assume that $I[\Phi, \Psi]$ is stationary for $\Phi = \Phi_0$ and $\Psi = \Psi_0$, and consider slightly deviated functions

$$\Phi(r) = \Phi_0(r) + \delta\eta(r), \tag{5A.100}$$

$$\Psi(r) = \Psi_0(r) + \delta\zeta(r), \tag{5A.101}$$

where $\eta(r)$ and $\zeta(r)$ are arbitrary continuous functions of r, and δ denotes a small real quantity. Substituting $\Phi(r)$ and $\Psi(r)$ into Eq. (5.242) and assuming that $I[\Phi, \Psi]$ is stationary for $\delta = 0$, we obtain

$$
\begin{aligned}
\frac{1}{2}\frac{\partial I}{\partial \delta}\bigg|_{\delta=0} = {} & \frac{1}{1-\chi}\int_0^a \frac{1-f}{\chi-f}\left[\frac{d\Phi_0}{dr}\frac{d\eta}{dr} + \frac{n^2}{r^2}\Phi_0\eta\right]r\,dr \\
& - \frac{k^2 n_1^2}{1-\chi}\int_0^a (1-f)\Phi_0\eta r\,dr \\
& + \int_0^a \frac{1}{\chi-f}\left[\frac{d\Psi_0}{dr}\frac{d\zeta}{dr} + \frac{n^2}{r^2}\Psi_0\zeta\right]r\,dr \\
& - k^2 n_1^2 \int_0^a \Psi_0\zeta r\,dr + \int_0^a \frac{n}{\chi-f}\frac{d}{dr}(\Phi_0\zeta + \Psi_0\eta)\,dr \\
& - \frac{1}{\chi-2\Delta}\bigg[\Omega_\beta \frac{1-2\Delta}{1-\chi}\Phi_0(a)\eta(a) \\
& \qquad + n\Phi_0(a)\zeta(a) + n\Psi_0(a)\eta(a) + \Omega_\beta\Psi_0(a)\zeta(a)\bigg] = 0. \quad (5A.102)
\end{aligned}
$$

By partial integration, Eq. (5A. 102) may be rewritten as

$$
\begin{aligned}
\frac{1}{2}\frac{\partial I}{\partial \delta}\bigg|_{\delta=0} = {} & -\int_0^a \frac{\eta(r)}{1-\chi}\left\{\frac{1}{r}\frac{d}{dr}\left[\frac{1-f}{\chi-f}r\frac{d\Phi_0}{dr}\right] + \left[k^2 n_1^2(\chi-f) - \frac{n^2}{r^2}\right]\right. \\
& \left. \times \frac{1-f}{\chi-f}\Phi_0 + \frac{n}{r}(1-\chi)\Psi_0\frac{d}{dr}\left(\frac{1}{\chi-f}\right)\right\}r\,dr \\
& - \int_0^a \zeta(r)\left\{\frac{1}{r}\frac{d}{dr}\left[\frac{1}{\chi-f}r\frac{d\Psi_0}{dr}\right] + \left[k^2 n_1^2(\chi-f) - \frac{n^2}{r^2}\right]\right. \\
& \left. \times \frac{1}{\chi-f}\Psi_0 + \frac{n}{r}\Phi_0\frac{d}{dr}\cdot\left(\frac{1}{\chi-f}\right)\right\}r\,dr \\
& + a\eta(a)\left\{\frac{1}{\chi-f(a)}\left[\frac{1-f}{1-\chi}\frac{d\Phi_0}{dr} + \frac{n}{r}\Psi_0\right]_{r=a}\right. \\
& \left. - \frac{1}{\chi-2\Delta}\left[\frac{1-2\Delta}{1-\chi}\frac{\Omega_\beta}{a}\Phi_0(a) + \frac{n}{a}\Psi_0(a)\right]\right\} \\
& + a\zeta(a)\left\{\frac{1}{\chi-f(a)}\left[\frac{d\Psi_0}{dr} + \frac{n}{r}\Phi_0\right]_{r=a}\right. \\
& \left. - \frac{1}{\chi-2\Delta}\left[\frac{\Omega_\beta}{a}\Psi_0(a) + \frac{n}{a}\Phi_0(a)\right]\right\}. \quad (5A.103)
\end{aligned}
$$

Since $\eta(r)$ and $\zeta(r)$ are arbitrary functions of r, this equation shows that $\Phi_0(r)$ and $\Psi_0(r)$ satisfy the vectorial wave equations to be solved [Eqs. (5.232) and (5.233)] and the boundary conditions at $r = a$:

$$\frac{1}{\chi - f(a)} \left[\frac{d\Psi_0}{dr} + \frac{n}{r} \Phi_0 \right]_{r=a} = \frac{1}{\chi - 2\Delta} \left[\frac{\Omega_\beta}{a} \Psi_0(a) + \frac{n}{a} \Phi_0(a) \right], \quad (5A.104)$$

$$\frac{1}{\chi - f(a)} \left[\frac{1 - f}{1 - \chi} \frac{d\Phi_0}{dr} + \frac{n}{r} \Psi_0 \right]_{r=a} = \frac{1}{\chi - 2\Delta} \left[\frac{1 - 2\Delta}{1 - \chi} \frac{\Omega_\beta}{a} \Phi_0(a) + \frac{n}{a} \Psi_0(a) \right],$$

$$(5A.105)$$

which give the continuity of the electric and magnetic fields, respectively. [The left-hand sides of Eqs. (5A.104) and (5A.105) express the fields in the core at $r = a$ (see Eqs. (5.237) and (5.239)). The right-hand sides express the fields in the cladding.]

5A.11   Proof that Eq. (5.258) Gives the Scalar Wave Approximations

We substitute the approximation under consideration [Eq. (5.258)] into Eqs. (5.232) and (5.233). Then, addition and subtraction of these equations yield

$$\frac{1}{r} \frac{d}{dr} \left[\frac{1}{x - f} r \frac{d\phi}{dr} \right] + \left[k^2 n_1^2 (\chi - f) - \frac{n^2}{r^2} \right] \frac{1}{x - f} \phi + \frac{n}{r} \phi \frac{d}{dr} \left(\frac{1}{x - f} \right) = 0,$$

$$(5A.106)$$

$$\frac{1}{r} \frac{d}{dr} \left[\frac{1}{\chi - f} r \frac{d\psi}{dr} \right] + \left[k^2 n_1^2 (\chi - f) - \frac{n^2}{r^2} \right] \frac{1}{\chi - f} \psi - \frac{n}{r} \psi \frac{d}{dr} \left(\frac{1}{\chi - f} \right) = 0,$$

$$(5A.107)$$

where

$$\phi(r) = \tfrac{1}{2} [\Phi(r) + \Psi(r)], \quad (5A.108)$$

$$\psi(r) = \tfrac{1}{2} [\Phi(r) - \Psi(r)]. \quad (5A.109)$$

Next we introduce the transverse field functions $R_{HE}(r)$ and $R_{EH}(r)$ defined as

$$R_{HE}(r) = \frac{1}{\chi - f}\left[\frac{d\phi}{dr} + \frac{n}{r}\phi\right], \tag{5A.110}$$

$$R_{EH}(r) = \frac{1}{\chi - f}\left[\frac{d\psi}{dr} - \frac{n}{r}\psi\right]. \tag{5A.111}$$

Substituting Eq. (5A.110) into (5A.106) and (5A.111) into (5A.107), we obtain

$$\frac{1}{r}\frac{d}{dr}\left(r\frac{dR_{HE}}{dr}\right) + \left[k^2n_1^2(\chi - f) - \frac{(n-1)^2}{r^2}\right]R_{HE} = 0, \tag{5A.112}$$

$$\frac{1}{r}\frac{d}{dr}\left(r\frac{dR_{EH}}{dr}\right) + \left[k^2n_1^2(\chi - f) - \frac{(n+1)^2}{r^2}\right]R_{EH} = 0, \tag{5A.113}$$

which are known as the scalar wave equations.

5A.12 Propagation Matrix Components [22]

(1) In those layers where $kn_i > \beta$,

$$p_{11} = \frac{\pi u_i a_i}{2}\left[J_n(u_i a_{i+1})N_{n-1}(u_i a_i) - N_n(u_i a_{i+1})J_{n-1}(u_i a_i)\right],$$

$$p_{12} = \frac{\pi u_i^2 a_i}{2\beta}\left[J_n(u_i a_{i+1})N_n(u_i a_i) - N_n(u_i a_{i+1})J_n(u_i a_i)\right],$$

$$p_{21} = \frac{\pi\beta a_i}{-2}\left[J_{n-1}(u_i a_{i+1})N_{n-1}(u_i a_i) - N_{n-1}(u_i a_{i+1})J_{n-1}(u_i a_i)\right],$$

$$p_{22} = \frac{\pi u_i a_i}{-2}\left[J_{n-1}(u_i a_{i+1})N_n(u_i a_i) - N_{n-1}(u_i a_{i+1})J_n(u_i a_i)\right], \tag{5A.114}$$

$$p_{33} = \frac{\pi u_i a_i}{-2}\left[J_n(u_i a_{i+1})N_{n+1}(u_i a_i) - N_n(u_i a_{i+1})J_{n+1}(u_i a_i)\right],$$

$$p_{34} = p_{12},$$

$$p_{43} = \frac{\pi\beta a_i}{-2}\left[J_{n+1}(u_i a_{i+1})N_{n+1}(u_i a_i) - N_{n+1}(u_i a_{i+1})J_{n+1}(u_i a_i)\right],$$

$$p_{44} = \frac{\pi u_i a_i}{2}\left[J_{n+1}(u_i a_{i+1})N_n(u_i a_i) - N_{n+1}(u_i a_{i+1})J_n(u_i a_i)\right],$$

where $u_i = (k^2 n_i^2 - \beta^2)^{1/2}$, and J_n and N_n denote the nth-order Bessel and Neumann functions, respectively.

(2) In those layers where $kn_i < \beta$,

$$p_{11} = w_i a_i [K_n(w_i a_{i+1}) I_{n-1}(w_i a_i) + I_n(w_i a_{i+1}) K_{n-1}(w_i a_i)],$$

$$p_{12} = \frac{w_i^2 a_i}{\beta} [-K_n(w_i a_{i+1}) I_n(w_i a_i) + I_n(w_i a_{i+1}) K_n(w_i a_i)],$$

$$p_{21} = \beta a_i [-K_{n-1}(w_i a_{i+1}) I_{n-1}(w_i a_i) + I_{n-1}(w_i a_{i+1}) K_{n-1}(w_i a_i)],$$

$$p_{22} = w_i a_i [K_{n-1}(w_i a_{i+1}) I_n(w_i a_i) + I_{n-1}(w_i a_{i+1}) K_n(w_i a_i)], \qquad (5A.115)$$

$$p_{33} = w_i a_i [K_n(w_i a_{i+1}) I_{n+1}(w_i a_i) + I_n(w_i a_{i+1}) K_{n+1}(w_i a_i)],$$

$$p_{34} = p_{12},$$

$$p_{43} = \beta a_i [-K_{n+1}(w_i a_{i+1}) I_{n+1}(w_i a_i) + I_{n+1}(w_i a_{i+1}) K_{n+1}(w_i a_i)],$$

$$p_{44} = w_i a_i [K_{n+1}(w_i a_{i+1}) I_n(w_i a_i) + I_{n+1}(w_i a_{i+1}) K_n(w_i a_i),$$

where $w_i = (\beta^2 - k^2 n_i^2)^{1/2}$, and I_n and K_n denote the nth-order modified Bessel functions of the first and second kinds, respectively.

8A.1 Mode Degeneration in a Quadratic-Profile Fiber

We follow Ref. [5] of Chapter 8. If the index distribution in a quadratic-profile fiber is expressed as

$$n^2(r) = n_1^2 [1 - (r/L)^2], \qquad (8A.1)$$

then the scalar wave equation for a field component R is

$$\frac{1}{r} \frac{\partial}{\partial r} \left(r \frac{\partial R}{\partial r} \right) + \frac{1}{r^2} \frac{\partial^2 R}{\partial \theta^2} + \frac{\partial^2 R}{\partial z^2} = -\omega^2 \varepsilon_0 \mu_0 n_1^2 \left[1 - \left(\frac{r}{L} \right)^2 \right] R. \qquad (8A.2)$$

Assuming a solution of the form

$$R = \psi(r) e^{-jm\theta} e^{-j\beta z}, \qquad (8A.3)$$

we obtain for $\psi(r)$

$$\frac{d^2 \psi}{dr^2} + \frac{1}{r} \frac{d\psi}{dr} + \left(-\frac{m^2}{r^2} + \sigma^2 - T^2 r^2 \right) \psi = 0, \qquad (8A.4)$$

where

$$\sigma^2 = \omega^2 \varepsilon_0 \mu_0 n_1^2 - \beta^2, \qquad (8A.5)$$

$$T = \omega(\varepsilon_0 \mu_0)^{1/2} n_1 / L. \qquad (8A.6)$$

The solution of Eq. (8A.4) is given as

$$\psi_{ml} = \left[\frac{T}{\pi}\frac{l!}{(l+m)!}\right]^{1/2} e^{-Tr^2/2}(Tr^2)^{m/2}L_l^{(m)}(Tr^2), \tag{8A.7}$$

where $L_l^{(m)}(x)$ are Laguerre's polynomials, and

$$\sigma^2 = \omega^2\varepsilon_0\mu_0n_1^2 - \beta^2 = \omega(\varepsilon_0\mu_0)^{1/2}n_1[2(2l+m+1)/L]. \tag{8A.8}$$

Hence, those modes having equal

$$m' = 2l + m \tag{8A.9}$$

degenerate with each other.

8A.2   Derivation of Eq. (8.27)

It is shown in Ref. [6] of Chapter 8 that when the relative index difference Δ is small, the delay time per unit distance of a uniform-core fiber is given as

$$dt/dz = (n_1/c)\Delta\{1 - (u/v)^2[1 - 2\xi_m(\omega)]\} + (n_1/c), \tag{8A.10}$$

where n_1 is the index of the core, c the light velocity in a vacuum, u and v the conventionally used parameters, and $\xi_m(\omega)$ is defined in Eq. (4.103). On the other hand, for those modes which are far from cutoff,

$$\xi_m(\omega) \simeq 1 - (1/v) \simeq 1 \tag{8A.11}$$

because $v \gg 1$. Therefore,

$$dt/dz \simeq (n_1/c)\Delta\{1 + (u/v)^2\} + (n_1/c). \tag{8A.12}$$

However, since

$$v = an_1k\sqrt{2\Delta}, \tag{8A.13}$$

$$u \simeq n_1k\theta a, \tag{8A.14}$$

where a is the core radius, k the wave number in a vacuum, and θ the quantity defined by Eq. (8.16), we obtain

$$\frac{dt}{dz} = \frac{n_1}{c}\Delta\left\{1 + \frac{\theta^2}{2\Delta}\right\} + \frac{n_1}{c} \simeq \frac{n_1}{c}\left(1 + \frac{\theta^2}{2}\right), \tag{8A.15}$$

which is the relation to be derived.

8A.3   Nature of Parameters Used in Eq. (8.36)

It is stated in Section 8.2.5 that parameters Θ_∞ [Eq. (8.37)] and γ_∞ [Eq. (8.38)] denote the width of the modal spread and the attenuation constant at the steady state $(z = \infty)$ under continuous light launching $(s = 0)$. Following Ref. [7] of Chapter 8, this fact will now be elucidated in more detail by using another solution of Eq. (8.30).

Letting $s = 0$ in Eq. (8.30), we obtain

$$\frac{\partial p}{\partial z} = -A\theta^2 p + \frac{D}{\theta}\frac{\partial}{\partial \theta}\left(\theta\frac{\partial p}{\partial \theta}\right). \tag{8A.16}$$

We assume here a solution of the form

$$p(\theta, z) = R(\theta)e^{-\gamma z}. \tag{8A.17}$$

Substitution of Eq. (8A.17) into (8A.16) yields

$$\frac{D}{\theta}\frac{\partial}{\partial \theta}\left(\theta\frac{\partial R}{\partial \theta}\right) = (A\theta^2 - \gamma)R, \tag{8A.18}$$

whose solutions are defined as Laguerre–Gauss functions R_n $(n = 0, 1, 2, \ldots)$. The attenuation constant is dependent upon n; i.e.,

$$R = \sum_{n=0}^{\infty} C_n R_n e^{-\gamma_n z} \tag{8A.19}$$

where the C_n are constants.

The attenuation constant γ_n increases with n. Therefore, for large z,

$$R \simeq C_0 R_0 e^{-\gamma_0 z}. \tag{8A.20}$$

This means that at long distances the field distribution R and the attenuation constant γ approach R_0 and γ_0, respectively, where

$$R_0(\theta) = \exp(-\theta^2/\Theta_\infty^2), \tag{8A.21}$$

$$\Theta_\infty = (4D/A)^{1/4}, \tag{8A.22}$$

$$\gamma_0 = (4DA)^{1/2}. \tag{8A.23}$$

The parameter γ_∞ in Eq. (8.38) is identical to γ_0 in Eq. (8A.23).

9A.1   The Induced Dipole Method

Scattering due to a small variation of the permittivity in a small volume in space can be computed by the "induced dipole method." This name originates from the fact that the scattered field is computed as the radiation from distributed equivalent dipoles corresponding to the spatial variation of the refractive index.

A small scattering region is represented by a sphere of radius a. The permittivities inside and outside this sphere are denoted by ε_1 and ε_2, respectively. When this sphere is immersed in a uniform static electric field $\mathbf{E}$, the effect of the presence of the sphere on the outside electric field is equal to that of a dipole having a moment

$$\mathbf{P}_s = 4\pi a^3 \frac{\varepsilon_2(\varepsilon_1 - \varepsilon_2)}{2\varepsilon_2 + \varepsilon_1} \mathbf{E} \tag{9A.1}$$

(see, e.g., W. K. H. Panofsky and M. Philips, "Classical Electricity and Magnetism," second edition, Addison-Wesley, Reading, Massachusetts, 1962).

The preceding statement can be extended to the following one; when a small volume V of any shape having permittivity ε_1 is present in the field $\mathbf{E}$, it is equivalent to a dipole having a moment

$$\mathbf{P} = \frac{3\varepsilon_2(\varepsilon_1 - \varepsilon_2)}{2\varepsilon_2 + \varepsilon_1} \mathbf{E}\, dV. \tag{9A.2}$$

Assuming that

$$\varepsilon_1 - \varepsilon_2 \ll \varepsilon_2, \tag{9A.3}$$

we can simplify Eq. (9A.2) to

$$\mathbf{P} = \Delta\varepsilon \mathbf{E}\, dV, \tag{9A.4}$$

where

$$\Delta\varepsilon = \varepsilon_1 - \varepsilon_2. \tag{9A.5}$$

On the other hand, an oscillating dipole

$$\mathbf{P}_0 = \mathbf{P} \exp j\omega t \tag{9A.6}$$

radiates an electromagnetic field whose electric field is expressed, on the coordinate system shown in Fig. 9A.1, as

$$\mathbf{E}_r = -(k^2 P/4\pi\varepsilon_2 r)e^{j\omega t}e^{-jkr}\sin\alpha \cdot \mathbf{a}, \tag{9A.7}$$

where $P = |\mathbf{P}|$, k is the wave number in the outside medium (ε_2), and $\mathbf{a}$ denotes a unit vector in the α direction.

We now assume that the small volume V with permittivity ε_1 is located at the origin of Fig. 9A.1 and a plane wave $\mathbf{E}$ polarized in the z direction is incident on it. In such a case, the scattered field $d\mathbf{E}_s$ is, from Eqs. (9A.4)

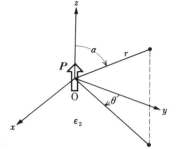

Fig. 9A.1 Coordinates and symbols used in the analysis.

and (9A.7),

$$dE_s = -(k^2/4\pi r)(\Delta\varepsilon/\varepsilon_2)Ee^{j\omega t}e^{-jkr}\sin\alpha \cdot \mathbf{a}\,dV, \tag{9A.8}$$

where $E = |\mathbf{E}|$.

Equation (9A.8) is the basic equation of the induced dipole method. When the permittivity variation in the scattering space is given, the scattered field can be computed by intergrating Eq. (9A.8) in that space.

9A.2 Sampling Theorem for Hankel Transform

We assume that functions $F(r)$ and $f(k)$ are related by Hankel and inverse-Hankel transform,

$$F(r) = \int_0^\infty kf(k)J_0(kr)\,dk, \tag{9A.9}$$

$$f(k) = \int_0^\infty rF(r)J_0(kr)\,dr. \tag{9A.10}$$

If we assume that

$$f(k) = 0 \qquad \text{for} \quad k > k_m, \tag{9A.11}$$

we may write

$$F(r) = \int_0^{k_m} kf(k)J_0(kr)\,dk = k_m^2 \int_0^1 \tilde{k}\tilde{f}(\tilde{k})J_0(k_m r\tilde{k})\,d\tilde{k}, \tag{9A.12}$$

where

$$k = k_m\tilde{k} \tag{9A.13}$$

and hence $\tilde{f}(\tilde{k}) = f(k_m\tilde{k})$.

If the zeros of $J_0(x)$ are denoted by α_{0l} ($l = 1, 2, \ldots$) and the function $\tilde{f}(\tilde{k})$ is finite and continuous in the region $[0, 1]$, the function is expressed by a Fourier-Bessel expansion,

$$\tilde{f}(\tilde{k}) = \sum_{l=1}^{\infty} a_l J_0(\alpha_{0l}\tilde{k}), \tag{9A.14}$$

where

$$a_l = \frac{2}{\{J_1(\alpha_{0l})\}^2} \int_0^1 t\tilde{f}(t) J_0(\alpha_{0l}t) \, dt. \tag{9A.15}$$

From Eqs. (9A.12) and (9A.15), we obtain

$$a_l = (2/\{J_1(\alpha_{0l})\}^2 k_m^2) F(\alpha_{0l}/k_m). \tag{9A.16}$$

Therefore,

$$f(k) = \sum_{l=1}^{\infty} (2/\{J_1(\alpha_{0l})\}^2 k_m^2) F(\alpha_{0l}/k_m) J_0((\alpha_{0l}/k_m)k). \tag{9A.17}$$

Thus, when a function $f(k)$ has values only in a limited range ($k < k_m$), it is uniquely determined by the values of its Hankel transform at discrete values $r = \alpha_{0l}/k_m$ ($l = 1, 2, \ldots$).

9A.3 Derivation of Eq. (9.50)

We follow Ref. [37] of Chapter 9. The ray equation is given in terms of cylindrical coordinates by Eqs. (3.17)–(3.19). From Eq. (3.18), we obtain

$$(d/ds)\{nr^2(d\phi/ds)\} = 0. \tag{9A.18}$$

If the ray is incident on the fiber at $(r_0, \phi_0, 0)$, then from Eqs. (3.19) and (9A.18),

$$n \, dz/ds = n(r_0) \, dz/ds|_0 \tag{9A.19}$$

$$nr^2 \, d\phi/ds = n(r_0) r_0^2 \, d\phi/ds|_0. \tag{9A.20}$$

Denoting the direction cosines of the ray before launch by $\cos \alpha_0$, $\cos \beta_0$, and $\cos \gamma_0$, we may rewrite Eq. (9A.19) as

$$n \, dz/ds = n(r_0) \cos \gamma_0$$

and hence,

$$n(d/ds) = n(r_0) \cos \gamma_0 \, d/dz. \tag{9A.21}$$

Multiplying Eq. (3.17) by n and combining with (9A.21), we obtain

$$n^2(r_0) \cos^2\gamma_0\{(d^2r/dz^2) - r(d\phi/dz)^2\} = n \, dn/dr = \tfrac{1}{2} d(n^2)/dr. \tag{9A.22}$$

Therefore, using $\varepsilon_s = n^2$,

$$(d^2r/dz^2) - r(d\phi/dz)^2 = [2\varepsilon_s(r_0)\cos^2\gamma_0]^{-1}\,d\varepsilon_s(r)/dr. \qquad (9A.23)$$

Since we assume in the text that $d\phi/dz = 0$ and $\varepsilon_s(r_0) = \varepsilon_{s1}$ we obtain Eq. (9.50) directly from Eq. (9A.23).

Index